Experiments in
Electric Circuit Analysis

Charles A. Schuler
California University of Pennsylvania
California, Pennsylvania

Richard J. Fowler
Professor Emeritus, Western Washington University
Bellingham, Washington

ILLUSTRATIONS BY
Jay D. Helsel
California University of Pennsylvania
California, Pennsylvania

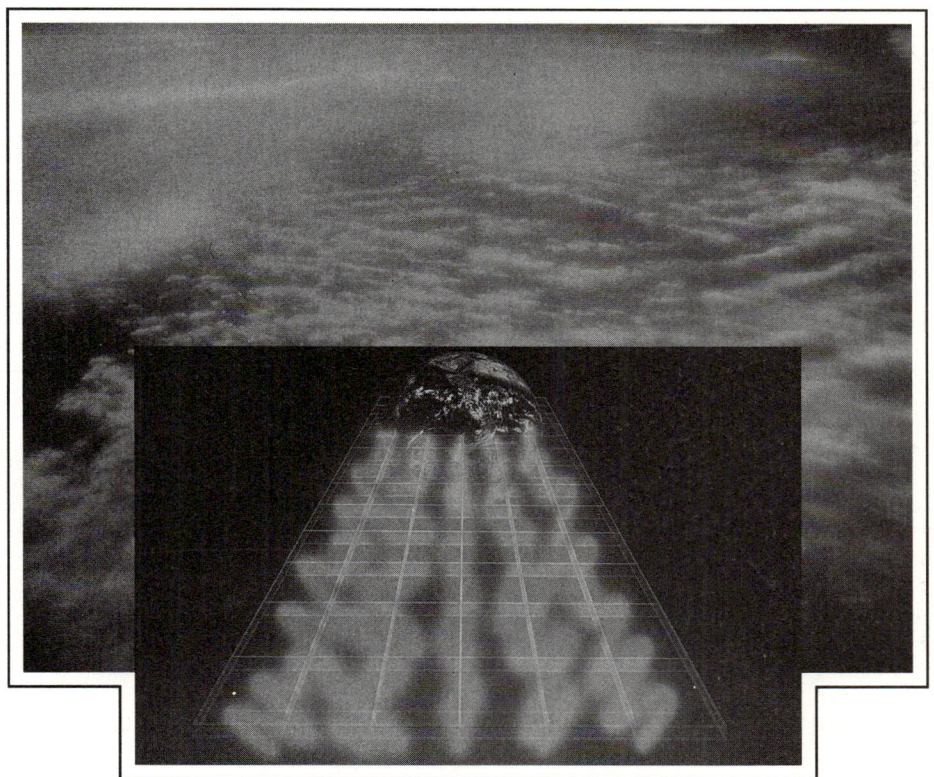

GLENCOE
Macmillan/McGraw-Hill

New York, New York   Columbus, Ohio   Mission Hills, California   Peoria, Illinois

Cover photographs:
Special Effects, Jook Leung/FPG International
Red Sky at Night, Telegraph Colour Library/FPG INTERNATIONAL
Lightning, Mike and Carol Werner/COMSTOCK, INC.

Experiments in Electric Circuit Analysis

Copyright © 1993 by the Glencoe Division of Macmillan/McGraw-Hill School Publishing Company. All rights reserved. Copyright © 1988 by McGraw-Hill, Inc., under the title *Experiments in Basic Electricity and Electronics*. All rights reserved. Except as permitted under the United States Copyright Act, no part of this publication may be reproduced or distributed in any form or by any means, or stored in a database or retrieval system, without the prior written permission of the publisher.

Send all inquiries to:
GLENCOE DIVISION
Macmillan/McGraw-Hill
936 Eastwind Drive
Westerville, OH 43081

ISBN 0-02-800443-4

Printed in the United States of America.

1 2 3 4 5 6 7 8 9   POH   01 99 98 97 96 95 94 93 92

Contents

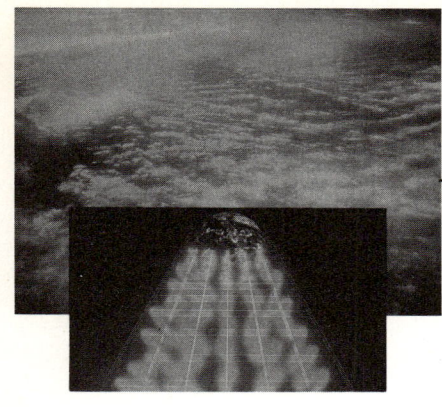

PREFACE		vi
CHAPTER 1	**Introductory Concepts**	**1**
ACTIVITY 1-1	Measurements and Conversions	1
CHAPTER 2	**Electrical Circuits and Quantities**	**5**
ACTIVITY 2-1	Circuit Measurements and Ohm's Law	5
CHAPTER 3	**Power and Energy**	**7**
ACTIVITY 3-1	The DC Power Supply	7
CHAPTER 4	**Electrical Materials**	**9**
ACTIVITY 4-1	Temperature and Resistance	9
CHAPTER 5	**Resistive Components**	**11**
ACTIVITY 5-1	Resistor Color Code	11
ACTIVITY 5-2	Potentiometer Characteristics	12
CHAPTER 6	**Series Circuits**	**15**
ACTIVITY 6-1	Series Circuit Concepts	15
ACTIVITY 6-2	Voltmeter Loading	17
ACTIVITY 6-3	Designing Series Circuits	17
CHAPTER 7	**Parallel Circuits**	**19**
ACTIVITY 7-1	Parallel Circuit Concepts	19
ACTIVITY 7-2	The Current Divider	20
ACTIVITY 7-3	Designing Parallel Circuits	21
CHAPTER 8	**Series-Parallel Circuits**	**23**
ACTIVITY 8-1	Series-Parallel Circuit Concepts	23
ACTIVITY 8-2	Voltage Dividers	24
ACTIVITY 8-3	The Wheatstone Bridge Circuit	24
ACTIVITY 8-4	Designing a Series-Parallel Circuit	25
CHAPTER 9	**Network Analysis**	**27**
ACTIVITY 9-1	Constant Current Source	27
ACTIVITY 9-2	Duals	28
ACTIVITY 9-3	Mesh Analysis	29
ACTIVITY 9-4	Superposition Theorem	30
ACTIVITY 9-5	Thevenin's Theorem	31
ACTIVITY 9-6	Norton's Theorem	32
ACTIVITY 9-7	Network Transforms	33
ACTIVITY 9-8	Analyzing an Unbalanced Bridge Circuit	35

CHAPTER 10 Magnetism — 37
- ACTIVITY 10-1 Permanent Magnets — 37
- ACTIVITY 10-2 Electromagnetism — 38

CHAPTER 11 Electrical Energy Sources — 41
- ACTIVITY 11-1 Electromagnetic Induction — 41
- ACTIVITY 11-2 Chemical Sources of Electricity — 42

CHAPTER 12 Alternating Current — 45
- ACTIVITY 12-1 Alternating Current Measurements — 45
- ACTIVITY 12-2 Phase — 46

CHAPTER 13 Inductance — 49
- ACTIVITY 13-1 Inductors and Mutual Inductance — 49

CHAPTER 14 Inductance in DC and AC Circuits — 53
- ACTIVITY 14-1 RL Time Constant — 53
- ACTIVITY 14-2 Inductors in AC Circuits — 54
- ACTIVITY 14-3 Inductive Reactance, Mutual Inductance and Coefficient of Coupling — 56
- ACTIVITY 14-4 Relationship of L, X_L, and f — 57

CHAPTER 15 Capacitance — 59
- ACTIVITY 15-1 Determining Capacitance — 59
- ACTIVITY 15-2 Capacitor Insulation Resistance — 60

CHAPTER 16 Capacitors in DC and AC Circuits — 63
- ACTIVITY 16-1 RC Time Constants — 63
- ACTIVITY 16-2 Capacitors in AC Circuits — 66
- ACTIVITY 16-3 Series-Parallel Capacitor Circuits — 68

CHAPTER 17 RCL Circuits — 69
- ACTIVITY 17-1 Differentiation and Integration — 69
- ACTIVITY 17-2 RCL (Impedance) Circuits — 72
- ACTIVITY 17-3 RC, RL, and RCL Circuits — 74

CHAPTER 18 Resonance and Filters — 77
- ACTIVITY 18-1 Resonant Circuits — 77
- ACTIVITY 18-2 Filters — 78
- ACTIVITY 18-3 Designing Filters — 82

CHAPTER 19 AC Network Analysis — 83
- ACTIVITY 19-1 Series-Parallel AC Circuits — 83
- ACTIVITY 19-2 Norton and Thevenin Equivalent AC Circuits — 84
- ACTIVITY 19-3 AC Network Problem — 86

CHAPTER 20 Transformers — 87
- ACTIVITY 20-1 Transformer Voltage Ratios and Phasing — 87
- ACTIVITY 20-2 Transformer Tests and Equivalent Circuits — 89
- ACTIVITY 20-3 Impedance Matching — 92
- ACTIVITY 20-4 Series-Aiding and Series-Opposing Windings — 95

CHAPTER 21	***Measuring Instruments***	***97***
ACTIVITY 21-1	Meter Tolerance and Loading	97
ACTIVITY 21-2	Shunts, Multipliers, and Ohmmeters	98
ACTIVITY 21-3	Oscilloscope Measurements	99
ACTIVITY 21-4	Designing an Analog Meter	101
CHAPTER 22	***Three-Phase Circuits***	***103***
ACTIVITY 22-1	Three-Phase Connections	103
ACTIVITY 22-2	Unbalanced Wye Loads	107
CHAPTER 23	***Introduction to Discrete Electronics***	***109***
ACTIVITY 23-1	Diode Characteristics	109
ACTIVITY 23-2	Rectifier Circuits	111
ACTIVITY 23-3	Transformer Internal Resistance	114
ACTIVITY 23-4	Filter Circuits	115
ACTIVITY 23-5	Zener Regulators	117
ACTIVITY 23-6	Low-Voltage Power Supply	118
CHAPTER 24	***Electronic Amplification***	***119***
ACTIVITY 24-1	Transistor Characteristics	119
ACTIVITY 24-2	Transistor Curves	122
ACTIVITY 24-3	BJT Amplifiers	125
ACTIVITY 24-4	FET Amplifiers	128
ACTIVITY 24-5	CE Amplifier Design	130
CHAPTER 25	***Integrated Circuits and Operational Amplifiers***	***131***
ACTIVITY 25-1	Differential Amplifier	131
ACTIVITY 25-2	Operational Amplifier	133
ACTIVITY 25-3	OP Amp Design	137
CHAPTER 26	***Introduction to Digital Circuits***	***139***
ACTIVITY 26-1	Decision-Making Gates	139
ACTIVITY 26-2	Flip-Flop Circuits	141
ACTIVITY 26-3	Implementing Boolean Expressions	145
MATERIALS LIST		**147**
APPENDIX 1	Activity Report	**149**
APPENDIX 2	BASIC Programs	**152**

Preface

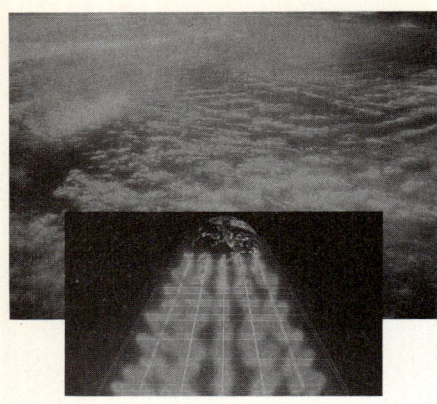

Laboratory activities and experiments are an essential part of a quality electricity/electronics program. This experiments manual provides a wide variety of laboratory experiences that have been tested and proven effective in a wide range of electricity/electronics programs.

The activities in this manual have a threefold purpose. First, they provide the student with experience in using general-purpose measuring devices to test and analyze circuits. Second, they require that the student learn to relate physical circuits to schematic diagrams and vice versa. Third, they reinforce the principles and concepts presented in the textbook and lectures by verifying (and using) basic laws, formulas, and electrical test procedures.

Although the manual presents activities appropriate for introductory courses that use any general electricity/electronics textbook, it was specifically designed and written to accompany *Electric Circuit Analysis*. The activities are presented in the same sequence as the topics in this textbook. The *Instructor's Manual* for the textbook provides an answer key, including typical experimental data, for the activities in this experiments manual.

The activities vary widely in level of difficulty and time required for completion. Also, there are more activities provided than can likely be used in a typical introductory program. Thus, the instructor can select those activities that best meet the objectives of a particular program.

Some of the activities in this manual are open-ended. They are structured to challenge the student to solve problems and use a more creative thought process. Appendix 1 of this manual contains a suggested format for the student reports for these open-ended activities.

The components, in the values needed for the experiments, are obtainable from most comprehensive electronics supply outlets. No special-purpose or sophisticated test equipment is needed.

Once a student has mastered the fundamental concepts of electrical/electronic circuits, the analysis and design of circuits can become a tedious exercise in mathematics. Therefore, many computer programs for analyzing circuits are presented in Appendix 2. These programs, which are written in BASIC for the IBM PC, take much of the tedium out of circuit analysis. The programs are also useful when students wish to check their solutions to complex circuits that they have solved by applying concepts and procedures that they are just learning. All of these programs are available on a diskette that accompanies the *Instructor's Manual*.

Charles A. Schuler
Richard J. Fowler

NAME _____   DATE _____

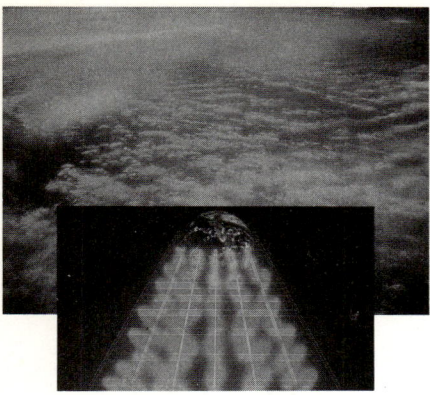

Chapter 1

Introductory Concepts

ACTIVITY 1-1 MEASUREMENTS AND CONVERSIONS

Introduction

The International System of Units (abbreviated SI) is a modernized version of the metric system of measurements. The system is built on a foundation of seven base units plus two supplementary units. Figure 1-1 shows the base and supplementary units and how they are used to derive additional SI units of measure. The base units for time, electric current, amount of substance, and luminous intensity are the same in both the U.S. Customary and SI systems.

Technical workers are confronted with several systems of measurement. Many of the U.S. Customary units are not the same as those in the SI system and unit conversions are often required. This activity is designed to provide experience with some of the measurements used in electricity and electronics and with conversions from one system to another.

Supplies

One or more electronics parts catalogs

Procedure

1. Physical sizes
   a. Search the catalog(s) for the physical sizes of various parts and components such as relays, integrated circuits, connectors, capacitors, circuit boards, switches, racks, lamps, batteries, displays, and transformers. What system of measurement is most often used? Are there inconsistencies; for example, can you find both fractional inch and decimal inch dimensions? Use Table 1-1 and convert a sample of U.S. Customary sizes to metric.

2. Temperature
   a. Use the catalog(s) and determine the temperature ratings for various components such as integrated circuits, capacitors, and crystals. What system of measurement is most often used? Is this consistent with your findings in part 1? Use Table 1-1 and convert a sample of the temperature ratings to U.S. Customary units.

3. Sensors
   a. Use the catalog(s) to find the ratings of various sensors such as strain gages, phototransistors, photodiodes, pressure transducers, accelerometers, and temperature sensors. Is there a mix among various systems of measurement? Use Table 1-1 to make the appropriate conversions from one system to another.

TABLE 1-1 Selected Unit Conversions and Definitions

Length
1 inch = 2.54 centimeters
1 micrometer = 1×10^{-6} meter
1 angstrom = 1×10^{-10} meter
1 mil = 1×10^{-3} inch

Volume
1 gallon = 3.785×10^3 cubic centimeters
1 barrel (U.S.) = 42 gallons
1 liter = 10^3 cubic centimeters

Force
1 pound = 4.448 newtons
1 dyne = 1×10^{-5} newton
1 kilogram = 9.807 newtons

Velocity and Acceleration
1 mile per hour = 1.609 kilometers per hour
1 foot per second = 30.48 centimeters per second
gravitational constant (g) = 9.807 meters per second^2

Copyright © 1993 by the Glencoe Division of Macmillan/McGraw-Hill School Publishing Company. All rights reserved.

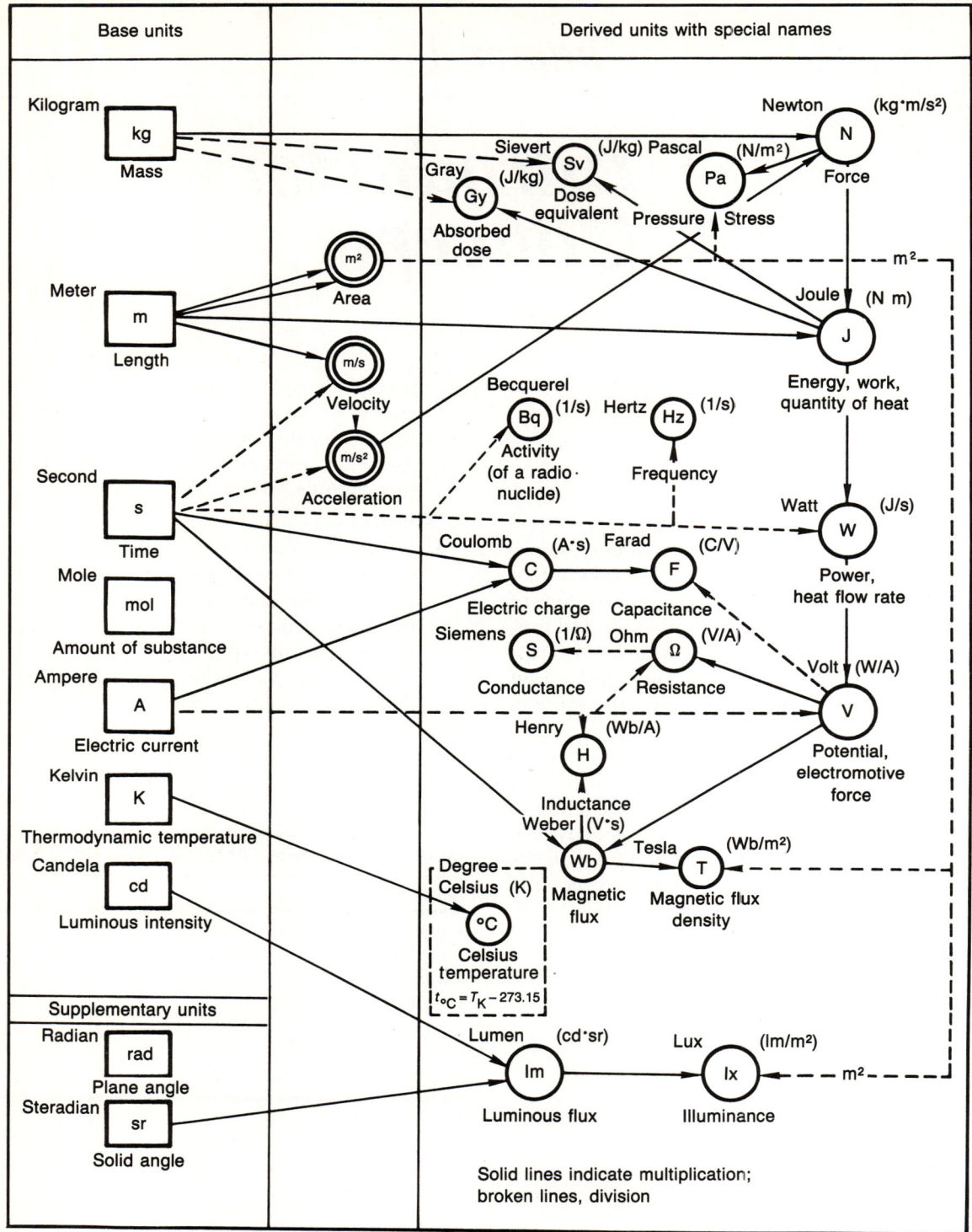

FIGURE 1-1 Relationships of SI units.

Table 1-1 Continued

Pressure

1 pound per inch^2 = 6.895 × 10^3 newtons per meter^2
1 dyne per centimeter^2 = 0.1 newton per meter^2
1 inch of mercury = 3.368 × 10^3 newtons per meter^2
atmospheric pressure (atm) = 1.013 × 10^5 newtons per meter^2
1 bar = 0.1 newton per centimeter^2
1 pascal = 1 newton per meter^2

Work and Energy

1 foot-pound = 1.356 joules
1 watt-hour = 3.6 × 10^3 joules
1 British thermal unit = 1.055 × 10^3 joules
1 calorie = 4.187 joules
1 kilogram-meter = 9.087 joules
1 erg = 1 dyne-centimeter = 1 × 10^{-7} joules

Power

1 foot-pound/second = 1.356 joules/second
1 horsepower = 745.7 joules/second
1 kilowatt = 1 × 10^3 joules/second
1 British thermal unit/hour = 0.2931 joules/second
1 ton (refrigeration) = 1.2 × 10^4 British thermal units
1 calorie/second = 14.29 British thermal units

Temperature

°F = 9/5°C + 32
°C = K − 273.15

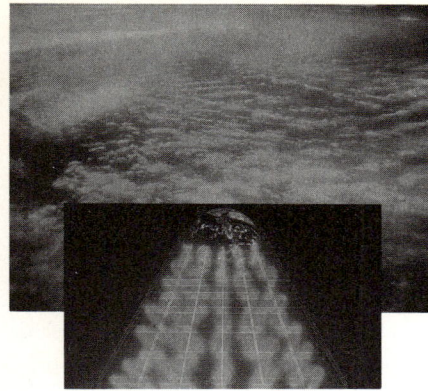

Chapter 2

Electrical Circuits and Quantities

ACTIVITY 2-1 CIRCUIT MEASUREMENTS AND OHM'S LAW

Introduction

One of the more fundamental skills in electricity and electronics is the ability to use instruments properly. This skill is essential when developing, verifying, adjusting, or troubleshooting circuits and equipment. Measurements are meaningful when they are compared to the values predicted by circuit laws. This activity deals with basic measurements and Ohm's law.

Supplies

(2) D cells, alkaline or carbon-zinc
(2) Resistors, 220-Ω, ½-W, ±5% (red-red-brown-gold color bands)
(1) DMM
(1) VOM

Procedure

1. Measuring resistance
   a. Use the DMM and select the resistance function and the 2-kΩ range. Measure the resistors. Each should show a resistance no lower than 209 Ω and no higher than 231 Ω. If one is out of this range, ask your instructor for a new resistor. If both are out of this range, have your instructor check your meter. Select the 200-Ω range and remeasure the resistors. What does the meter indicate? Select the 20-kΩ range and remeasure the resistors. Is this the range for best resolution? Why?
   b. Use the VOM and select the resistance function and the $R \times 1$ range. Make sure the pointer is at the leftmost calibration mark with the meter probes separated. Touch the probes together and adjust the ohm's control if necessary for a reading of 0. Measure the resistors. Is this the range for best resolution? Select a range that produces a near midscale reading. Recheck the zero by shorting the probes and then remeasure the resistors. Assuming both meters are working properly, which do you think is the most accurate? Which is the easiest to use in this application?
   c. Use either meter and measure the resistance with the resistors connected as shown in Fig. 2-1. This is an example of what often happens during in-circuit testing when using an ohmmeter and why it may be necessary to isolate the component being measured. When measuring the resistance of a component in a circuit, a _____ than normal reading may occur.

2. Measuring dc voltages
   a. Set the DMM for a 2-V dc range. Measure the cell voltage as shown in Fig. 2-2(a). Note that the "button" end of the cell is positive. What is the cell voltage?
   b. Connect the cells as shown in Fig. 2-2(b). Select a higher range and measure the voltage. How does this reading compare with step 2a?
   c. Connect the cells as shown in Fig. 2-2(c). What is the dc voltage now?

FIGURE 2-1 Simulating in-circuit resistance measurement.

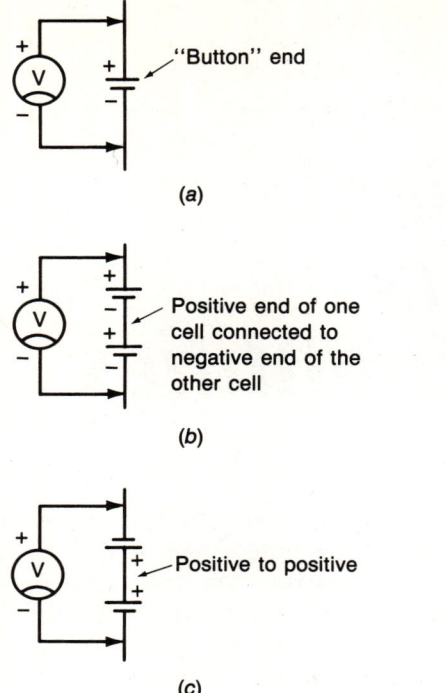

FIGURE 2-2 Voltage measurements.

d. Select the dc voltage range for best resolution on the VOM. Measure cell voltage. Which meter should be more accurate in this application?

3. Measuring dc current
   a. Use Ohm's law and calculate the current flow for the circuit shown in Fig. 2-3(a). Build the circuit. Set the DMM for dc current and select a 20-mA range. Break the circuit at some point and insert the meter. Observe polarity. The measured current should be within a few percent of the calculated current. If it is significantly lower, check the cell voltage under load (with the 220-Ω resistor connected). An exhausted cell will show a drop in output when loaded. Reverse the meter polarity. Does the display indicate a minus sign? How would the VOM react if connected this way?
   b. Use Ohm's law and calculate the current for the circuit shown in Fig. 2-3(b). Build the circuit. Measure the current using the DMM. Doubling the voltage in a circuit will cause the current to _____.
   c. Build the circuit shown in Fig. 2-3(c) and measure the current. Compare this current to that measured in step 3a. Doubling the resistance of a circuit will cause the current to _____.
   d. Verify that the resistance is doubled as shown in Fig. 2-3(d).

⇨ **CAUTION:** Make sure that the cell is removed from the circuit before connecting the ohmmeter.

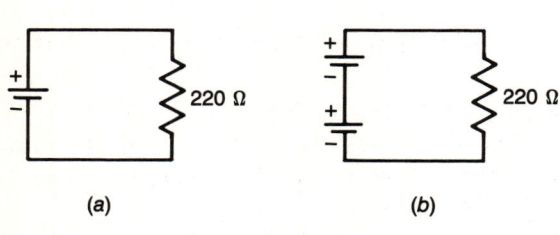
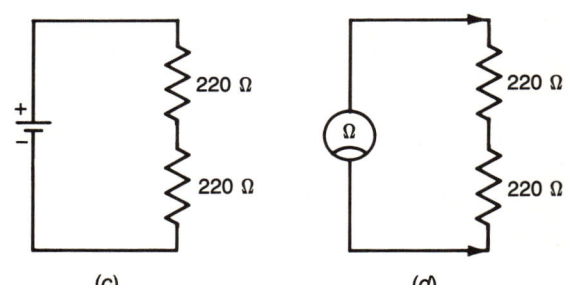

FIGURE 2-3 Circuits for current measurements.

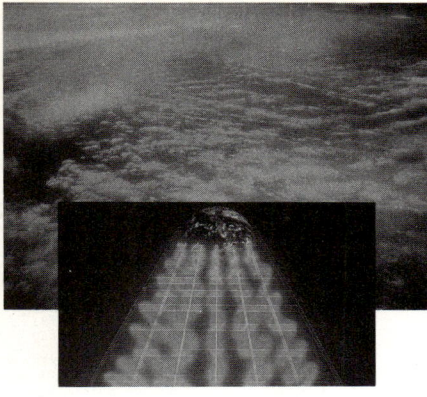

Chapter 3

Power and Energy

ACTIVITY 3-1 THE DC POWER SUPPLY

Introduction

Cells and batteries are convenient sources of electrical energy for many applications. However, they are often not desirable for circuit testing and evaluation. DC power supplies are common items on almost all electrical and electronic test benches. They usually offer adjustable voltage output and many also include adjustable current limiting. Technicians must know the proper operating procedures for bench power supplies.

Supplies

(1) Resistor, 100-Ω, ½-W, ±5% (brown-black-brown-gold color bands)
(1) Resistor, 100-Ω, 2-W, ±5% (brown-black-brown-gold color bands)
(1) Power supply, 0- to 25-V dc (with adjustable current limiting)
(1) DMM or VOM

Procedure

1. Voltage adjust
   a. Set your DMM for a dc voltage range that is at least as large as the maximum voltage output rating for the supply. Connect the meter leads to the supply output terminals. Turn on the supply. Adjust the supply output for zero volts. If the supply has an internal meter, it should read zero and so should the external meter. Now, adjust the supply voltage to about midrange. Check for reasonable agreement from one meter to the other. Finally, check the supply at maximum voltage output. Any abnormality in supply or meter operation should be reported to your instructor. Turn the supply off.

2. Current limiting
   a. Calculate the amount of current that will produce 2-W of dissipation in a 100-Ω resistor. This may be considered the maximum safe current since more would cause a power dissipation in excess of the resistor's rating.
   b. Build the circuit shown in Fig. 3-1. Do not turn the power supply on yet. Note that the supply is represented by the symbol for an adjustable battery. Use Ohm's law or the power formula to calculate the maximum safe voltage across the resistor. Set the meter for a 200-mA dc range.
   c. Adjust the current limit control on the power supply to midrange. Adjust the voltage control to zero. Turn on the supply. Adjust the voltage to the value calculated in the previous step. You may use the supply internal meter or a voltmeter connected across the supply output terminals. If the desired voltage cannot be reached, you may have to increase the setting of the current control. *Cautiously* touch the resistor. It should become very warm. What does the current meter read? What is the resistor dissipation?
   d. Slowly turn down the current limit control while watching the current meter. A point should be reached where the meter reading starts to drop. Leave the control at that point. Now, increase the setting of the voltage control. Note that the current does not increase. Set the voltage control to maximum. Again, note that the current

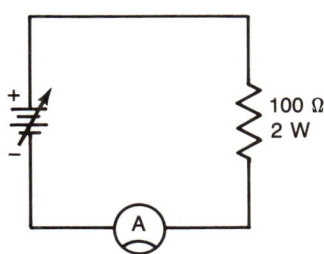

FIGURE 3-1 Experimenting with current limiting.

Copyright © 1993 by the Glencoe Division of Macmillan/McGraw-Hill School Publishing Company. All rights reserved.

does not increase. Notify your instructor if you cannot obtain the correct response.

e. Do not change the setting of the current control. Set the voltage control to zero. Slowly increase the output voltage while watching the current. Note that the current increases as a linear function of the voltage. What happens when the current limit is reached?

f. *CAUTION: Do not perform this step if the current limit on your supply is not working properly.* Do not change the setting of the current control. Set the voltage control to zero. Remove the resistor from the circuit as shown in Fig. 3-2. Slowly increase the output while watching the current. Does the supply go into current limiting almost immediately? What could happen in this case if the supply is not current-limited? Turn off the supply.

3. Thermal mass and temperature rise
   a. Calculate the maximum safe voltage across a 100-Ω, ½-W resistor.
   b. Build the circuit shown in Fig. 3-3. Set the supply for the output voltage calculated in the previous step. Use a voltmeter and verify that the voltage is the same across both resistors. Since

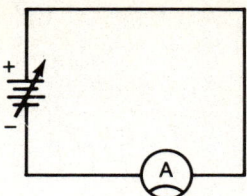

FIGURE 3-2 Shorting a current-limited supply.

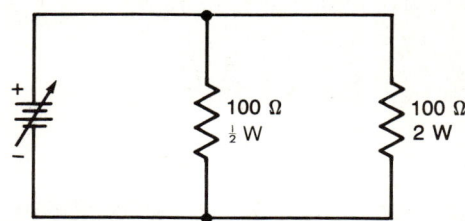

FIGURE 3-3 Comparing temperature rise in two resistors.

the resistors are equal in value, what can you conclude about the power dissipation?

c. After about 3 min, check the temperature of each resistor. Are they equal in temperature? Why?

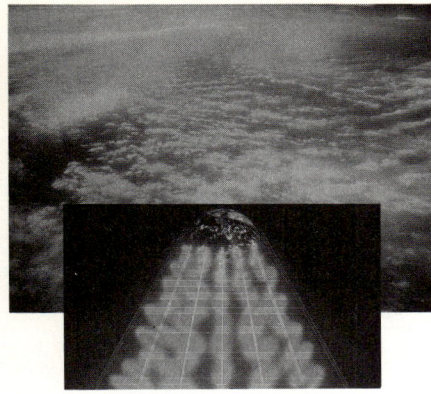

Chapter 4

Electrical Materials

ACTIVITY 4-1 TEMPERATURE AND RESISTANCE

Introduction

The temperature response of electrical materials and devices is of major importance in many applications. Some materials show an increase in opposition to current flow as temperature increases. Some show a decrease and others show very little change.

Supplies

(1) Resistor, 100-Ω, 2-W, ±5% (brown-black-brown-gold color bands)
(1) Lamp, 14-V, 0.08-A, no. 756, with holder
(2) Transistors, 2N4124, or equivalent
(1) Soldering pencil, 23-W
(1) VOM
(1) DMM
(1) Power supply, 0 to 25-V dc

Procedure

1. Cold lamp resistance
   a. Use your DMM or VOM and measure the resistance of the lamp.
   b. The lamp is rated at 14 V. Use Ohm's law to calculate the lamp current using its rated voltage and the resistance value measured in the previous step.
   c. The lamp is rated at 0.08 A. Use its rated voltage and rated current to calculate its resistance.
   d. How does the resistance calculated in the previous step compare with the resistance measured in the first step? Why?
2. Lamp characteristic curve
   a. Build the circuit shown in Fig. 4-1. The current meter should be set on a 200-mA range and the voltmeter on a 15-V range. CAUTION: Do not apply more than 14 V to the lamp. Start at 0 V and record the current. Increase the power supply in 1-V intervals and record the current. Plot your data on the graph provided in Fig. 4-2.
   b. Is the lamp volt-ampere characteristic curve linear (a straight line)? Why?
3. Resistor characteristic curve
   a. Turn off the power supply and remove the lamp from the circuit. Replace it with a 100-Ω, 2-W resistor. Do not set the power supply for more than 14 V. Collect data in 1-V intervals and plot the resistor characteristic curve on the graph provided in Fig. 4-2. Is the resistor a linear device? Is the lamp a linear device?
4. Semiconductor temperature response
   a. Plug in the soldering pencil. Place it in a position where it is out of the way. Allow several minutes for it to reach operating temperature.
   b. Build the circuit shown in Fig. 4-3. Set the ohmmeter on its highest range. It should indicate infinity ohms or overrange when connected as shown. If it does not, check the ohmmeter polarity and circuit connections.
   c. Apply heat to the C lead of transistor Q_1. The tip of the soldering pencil should be about $\frac{1}{8}$ in from the case of Q_1. Hold the tip to the lead for 20 s. What does the ohmmeter indicate? What does the ohmmeter do when the heat is removed from the transistor?

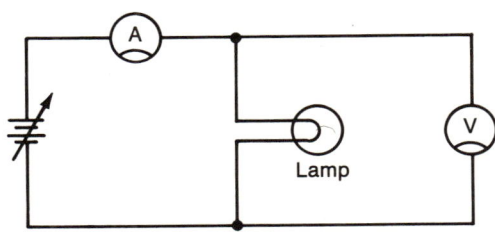

FIGURE 4-1 Circuit for measuring lamp characteristics.

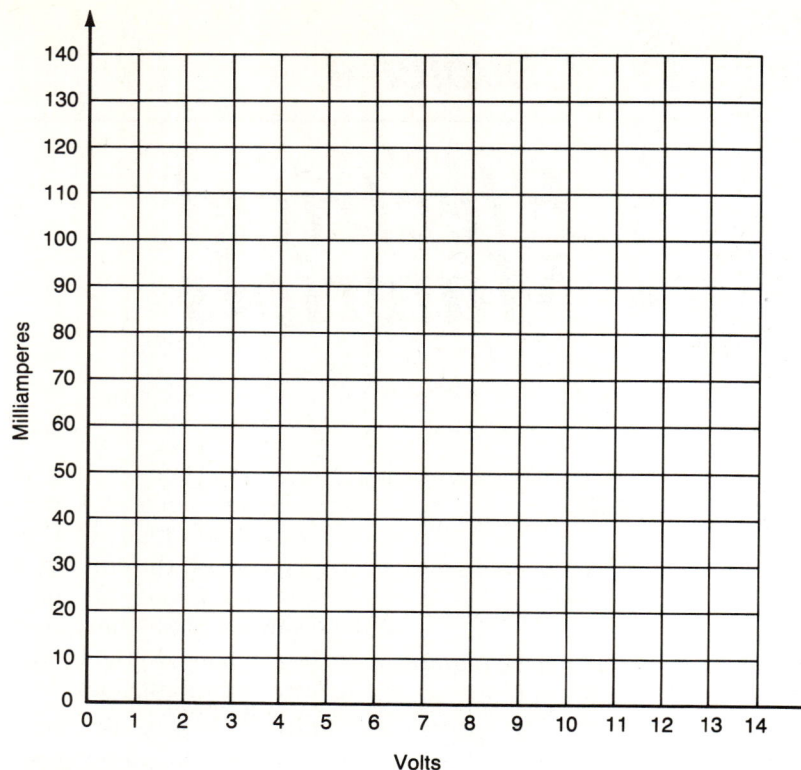

FIGURE 4-2 Lamp characteristic graph.

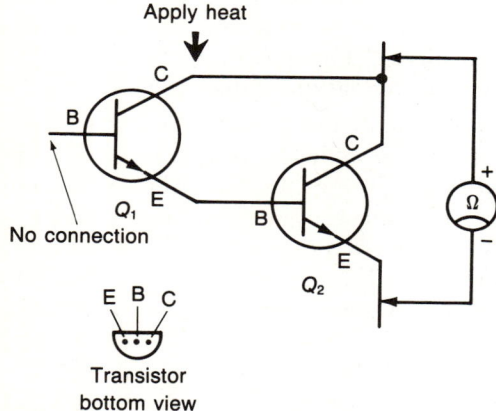

FIGURE 4-3 Semiconductor temperature response circuit.

d. The transistors are made from silicon, which is a semiconductor material. What type of temperature coefficient is indicated by your results? The lamp filament is made from a tungsten alloy, which is a metal. What type of temperature coefficient is indicated by your results? The resistor characteristic curve is a straight line. What type of temperature coefficient does this indicate?

NAME _____ DATE _____

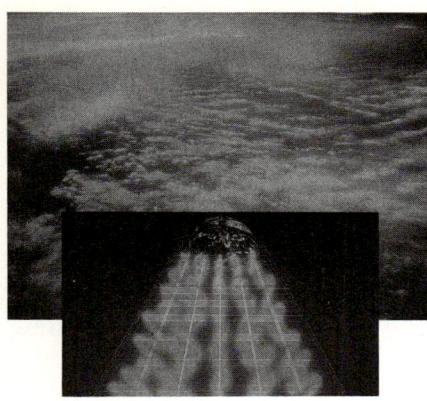

Chapter 5

Resistive Components

ACTIVITY 5-1 RESISTOR COLOR CODE

Introduction

Resistors with dissipation ratings ranging from $\frac{1}{8}$- to 2-W are generally coded with color bands. The bands indicate the nominal resistance value and the tolerance of the resistor. As you work with the color code you will become efficient in its use.

Supplies

(1) Assortment of $\frac{1}{2}$-W, ±5% resistors (values shown in Table 5-1)
(1) DMM or VOM

Procedure

1. Identifying and measuring resistors
   a. Use your meter and your knowledge of the color code to complete Table 5-1. Report out-of-tolerance resistors to your instructor. If quite a few appear to be out of tolerance, check your meter for proper operation.

TABLE 5-1 Data for Resistor Color Codes and Measurements

Nominal Value	Color Bands				Tolerance Range	Measured Value	Within Tolerance
	1st	2nd	3rd	4th			
1 Ω							
4.7 Ω							
10 Ω							
27 Ω							
100 Ω							
220 Ω							
560 Ω							
1 k Ω							
5.1 k Ω							
10 k Ω							
22 k Ω							
68 k Ω							
120 k Ω							
150 k Ω							
180 k Ω							
910 k Ω							

Copyright © 1993 by the Glencoe Division of Macmillan/McGraw-Hill School Publishing Company. All rights reserved.

ACTIVITY 5-2 POTENTIOMETER CHARACTERISTICS

Introduction

Potentiometers are the most widely applied type of variable resistor. They have three terminals and a sliding contact called the "wiper arm." The wiper arm is usually connected to the center terminal.

Supplies

(1) Potentiometer, 1-kΩ, $\frac{1}{2}$-W
(1) DMM or VOM
(1) Power supply, 0- to 25-V dc

Procedure

1. Measuring resistance versus rotation
   a. Figure 5-1 shows the orientation of a typical potentiometer. Use your multimeter to measure its resistance. Choose a range that gives good resolution at 1 kΩ. Measure the resistance from terminal 1 to terminal 3. It should be between 900 and 1100 Ω. Rotate the shaft and note that the resistance from terminal 1 to terminal 3 does *not* change.
   b. If the shaft of your potentiometer has a slotted end, use it to determine the percentage of shaft rotation. If there is no slotted end, place a reference mark on the end of the shaft with a pencil or a pen.
   c. Rotate the shaft fully counterclockwise (ccw), and mark this shaft position on the body of the potentiometer as 0 percent. Rotate the shaft fully clockwise (cw), and mark this shaft position as 100 percent. Make additional calibration marks between 0 and 100 by visually subdividing the arc lengths.
   d. Connect the ohmmeter across terminals 2 and 3 of the potentiometer. Start at full ccw (0 percent), and measure the resistance. Record the results, and then rotate the shaft to the next calibration. Collect sufficient data to complete the graph shown in Fig. 5-2.
2. Measuring output voltage versus rotation
   a. Set your dc power supply for 10 V output. Set your multimeter to dc volts. Choose a range that gives good resolution for values up to 10 V.
   b. Connect the circuit shown in Fig. 5-3.
   c. Start at full ccw (0 percent), and measure the output voltage with your voltmeter. Record the results, and then rotate the shaft to the next calibration mark. Collect sufficient data to complete the graph shown in Fig. 5-4.

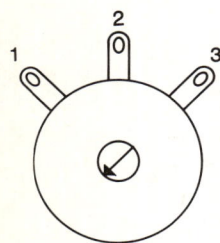 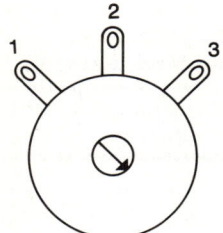

Full ccw (0%)   Full cw (100%)

FIGURE 5-1 Potentiometer orientation.

NAME _____   DATE _____

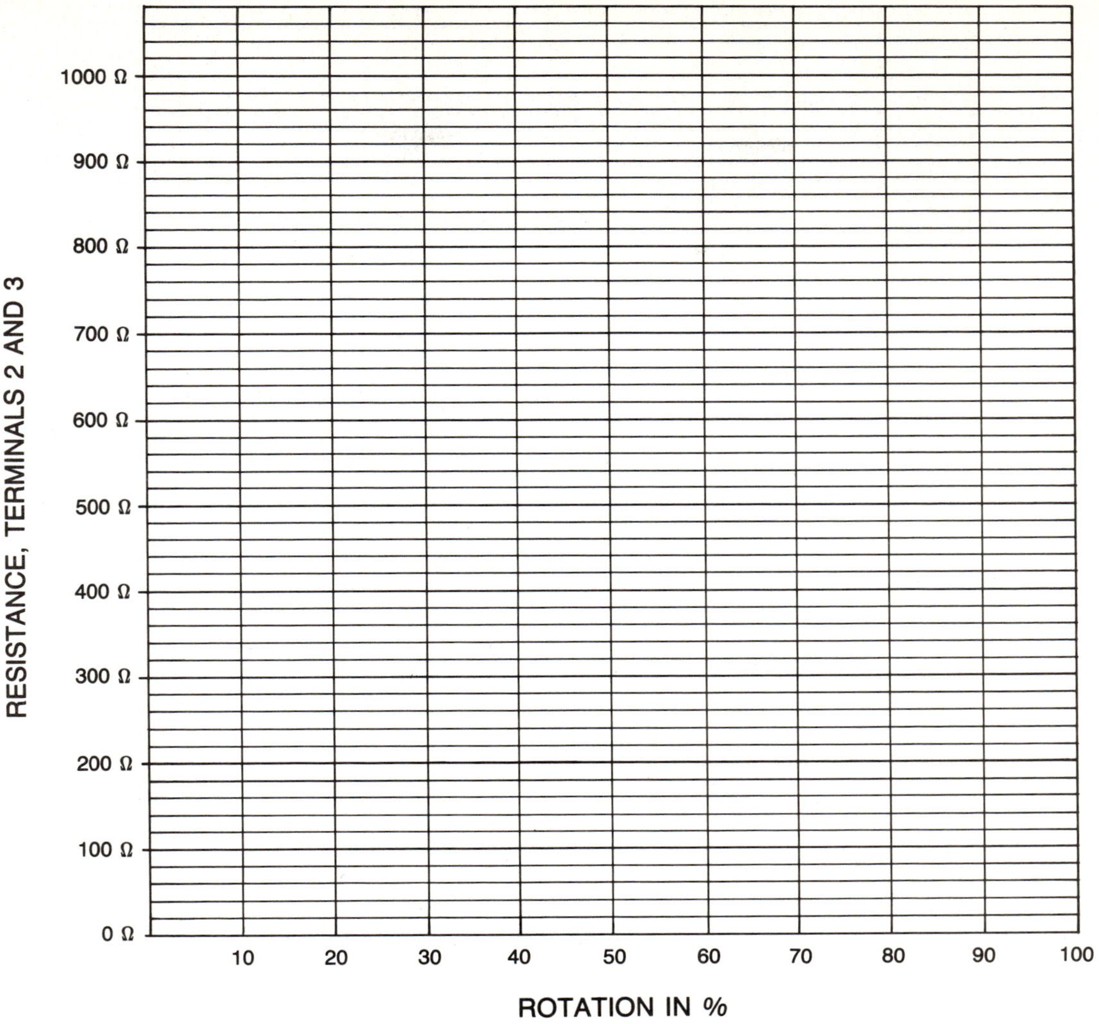

FIGURE 5-2 Resistance versus rotation.

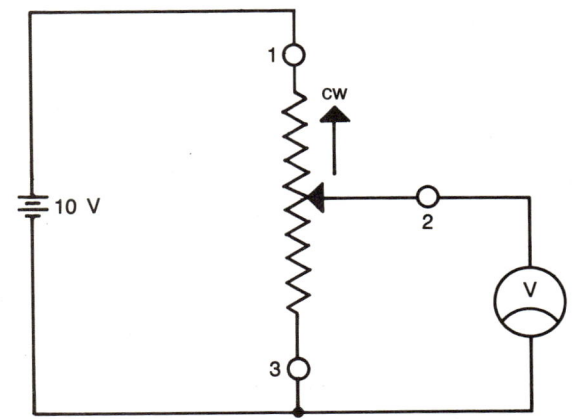

FIGURE 5-3 Potentiometer circuit.

Copyright © 1993 by the Glencoe Division of Macmillan/McGraw-Hill School Publishing Company. All rights reserved.

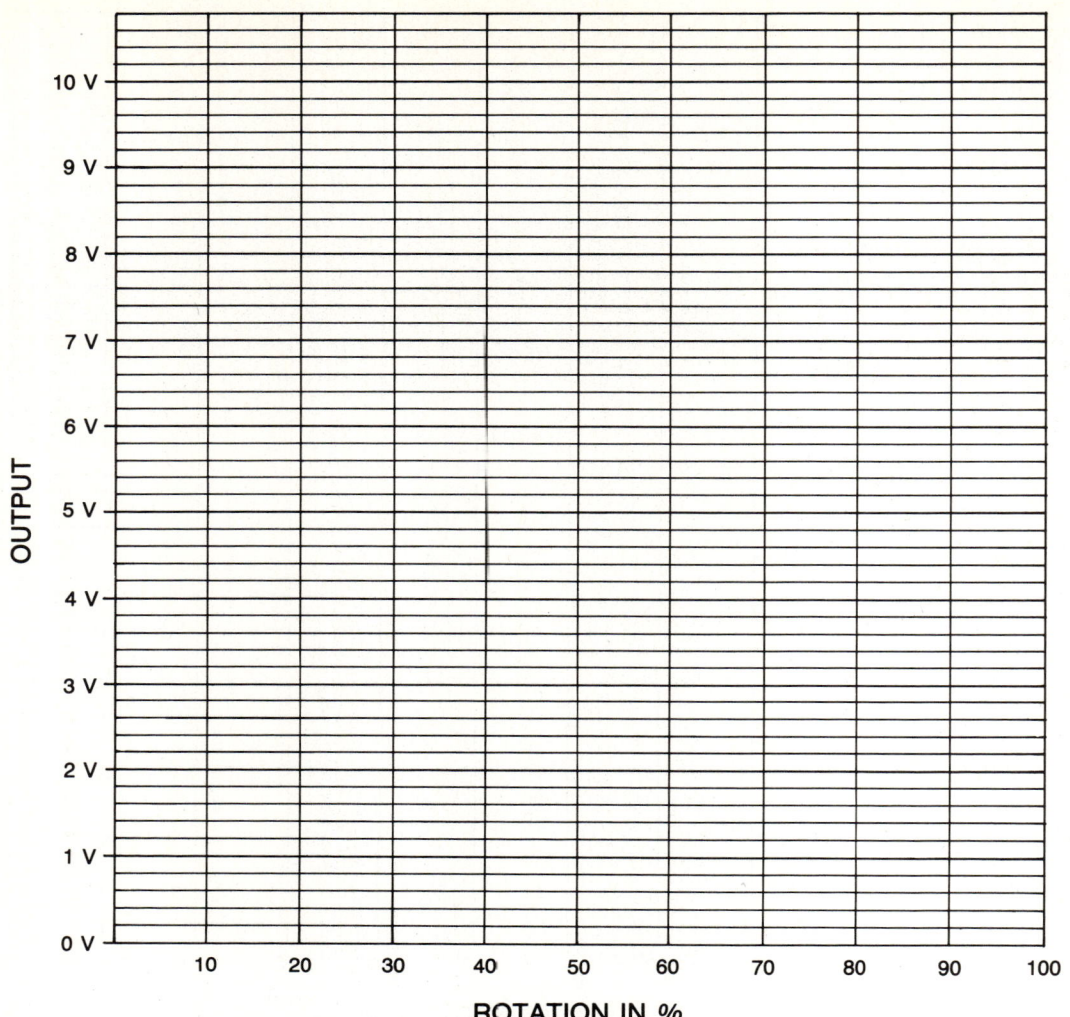

FIGURE 5-4 Output voltage versus rotation.

NAME _____ DATE _____

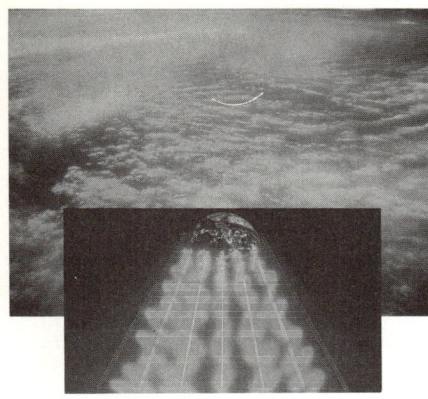

Chapter 6

Series Circuits

ACTIVITY 6-1   SERIES CIRCUIT CONCEPTS

Introduction

Series circuits behave according to Ohm's law and Kirchhoff's voltage law. These laws are critical for understanding all electrical circuits. This activity investigates the laws for series circuits.

Supplies

(3)   Resistors, 220-Ω, ½-W, ±5%
(1)   Resistor, 10-kΩ, ½-W, ±5%
(1)   Potentiometer, 1-kΩ, ½-W
(1)   Lamp, 14-V, 0.08-A, no. 756, with holder
(1)   DMM or VOM
(1)   Power supply, 0- to 25-V dc

Procedure

1. Series circuit concepts
   a. Connect three 220-Ω resistors in series as shown in Fig. 6-1. Do not connect the power supply yet. Calculate and measure the total resistance. Record your results in the table provided in Fig. 6-2. Calculate all of the remaining values shown in the table. Connect the power supply and measure all of the values except power.
   b. Measure the current by breaking into the circuit at point A. Then measure the current at points B, C, and D. What is constant in a series circuit?
   c. What can you conclude about the voltage drops when equal value resistors are in series?
   d. If point C in Fig. 6-1 is used as a reference point, what is the voltage at point B? At point D?
   e. Change the circuit as shown in Fig. 6-3. Measure total resistance before connecting the power supply. Complete the table given in Fig. 6-4.
   f. When a series circuit contains one resistor that is much higher in value than the other resistors in

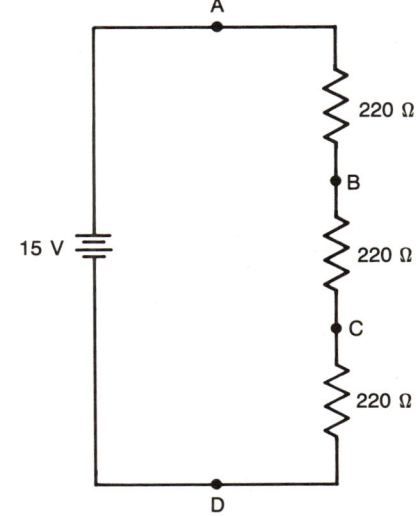

FIGURE 6-1   Series circuit.

the circuit, what can you conclude about the drop across the large resistance?

2. Potentiometers as voltage dividers
   a. Build the circuit shown in Fig. 6-5. Adjust the potentiometer while observing the voltmeter. What range of voltage is available from the divider?

	Calculated	Measured
R_T		
I		
V_{AB}		
V_{BC}		
V_{AC}		
V_{CD}		
V_{BD}		
P_T		

FIGURE 6-2   Data table.

Copyright © 1993 by the Glencoe Division of Macmillan/McGraw-Hill School Publishing Company. All rights reserved.

15

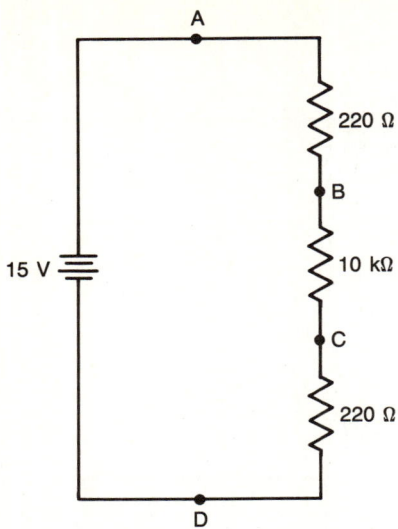

FIGURE 6-3 Modified series circuit.

b. Observe whether the meter reading increases or decreases when the potentiometer control is rotated clockwise. Change the circuit shown in Fig. 6-5 so that the voltmeter is across the top terminal of the potentiometer and the wiper arm. Is it possible to wire a potentiometer for either increasing or decreasing output with clockwise rotation?

	Calculated	Measured
R_T		
I		
V_{AB}		
V_{BC}		
V_{AC}		
V_{CD}		
V_{BD}		
P_T		

FIGURE 6-4 Data table.

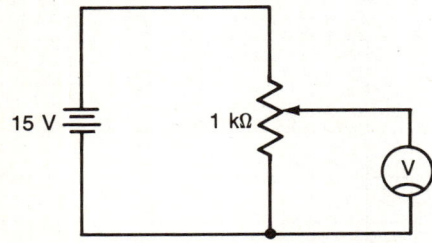

FIGURE 6-5 Potentiometer voltage divider.

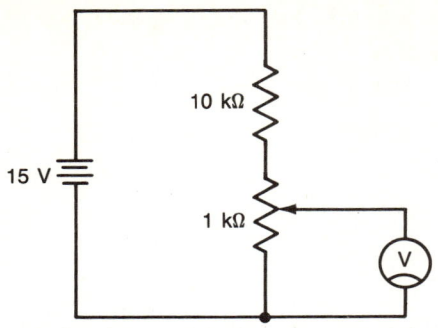

FIGURE 6-6 Modified potentiometer voltage divider.

c. Use a DMM if one is available. Attempt to adjust the potentiometer for an output of exactly 1 V. Is this adjustment "touchy"?

d. Build the circuit shown in Fig. 6-6. Once again attempt to adjust the output for exactly 1 V. Is the adjustment less touchy? What is the range of output voltage now?

3. Simulating series circuit faults
   a. Build the circuit shown in Fig. 6-7. Be sure to set the supply for 14 V. Is the lamp operating at normal output?
   b. Measure the drop across the lamp. Measure the drop across the resistor. What is the sum of the drops?
   c. Short the resistor with a test lead. What does the lamp show? What is the voltage drop across the lamp now?
   d. Remove the short and replace the 220-Ω resistor with a 10-kΩ resistor. What does the lamp show? Measure the drop across the resistor. Measure the drop across the lamp.
   e. What effect will a shorted component have on current flow in a series circuit? What happens to the power dissipation in the other components when one component shorts?
   f. What happens in a series circuit when one component opens or develops a very high resistance?

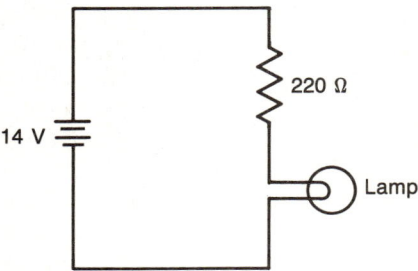

FIGURE 6-7 Simulating series circuit faults.

ACTIVITY 6-2 VOLTMETER LOADING

Introduction

Series circuit concepts explain some of the common errors in measurement that can occur when you are working with electrical devices and circuits. For example, the resistance of test leads and connections can cause an error when low values of resistance are measured. If the leads of an ohmmeter are touched together, the meter should ideally indicate 0 Ω. If it does not, and if the meter cannot be adjusted for zero (as in the case of a DMM), the indicated reading is the residual value which must be subtracted when a low value of resistance is measured. Suppose a DMM shows a residual value of 0.4 Ω. That value must be subtracted from the reading when low values of resistance are measured. If you didn't know about this, you might decide that a normal 1-Ω, ±10 percent resistor is out of tolerance because it would read 1.4 Ω. In fact, there is nothing wrong with the resistor, since the residual resistance is adding in series with the resistance under test. This activity demonstrates another series effect called "voltmeter loading."

Supplies

(1) Resistor, 1-kΩ, ½-W, ±5%
(1) Resistor, 910-kΩ, ½-W, ±5%
(1) DMM or VOM
(1) Power supply, 0- to 25-V dc

Procedure

1. Voltmeter loading
   a. Set your meter for the dc voltage range that gives the best resolution at 10 V.
   b. Set your power supply for 10 V dc.
   c. Connect the meter to the power supply and carefully readjust the supply so that the meter indicates as closely as possible to 10 V.
   d. Connect the supply and 910-kΩ resistor as shown in Fig. 6-8(a).
   e. Measure the voltage from A to B in Fig. 6-8(a). Is the reading what you expected? Can you explain the error?
   f. Figure 6-8(b) shows how to analyze the effect of voltmeter loading. If the meter has a resistance of 10 MΩ (as do most DMMs), the voltage divider equation can be used to find the drop from A to B.
   g. Use the voltage divider equation to verify that the voltage from A to B in Fig. 6-8(b) will be 9.17 V.
   h. If your meter has a resistance other than 10 MΩ, find out what it is and use its value to explain the error you obtained in step 1e.
   i. Remove the 910-kΩ resistor from your circuit and replace it with a 1-kΩ resistor. Measure the voltage from A to B again. Is there any error now? Why?

ACTIVITY 6-3 DESIGNING SERIES CIRCUITS

Introduction

The first two activities involving series circuits specified the circuit diagrams, component values, and source conditions to be used, as well as a step-by-step procedure to be followed. This activity uses a different approach. It requires you to do some very basic circuit design. It specifies the desired circuit characteristics, and you must design the circuit, select the components, and develop the procedure needed to verify your design experimentally.

Supplies

(1) Power supply, 0- to 25-V dc
(1) DMM or VOM

Miscellaneous components selected from the Materials List in this manual

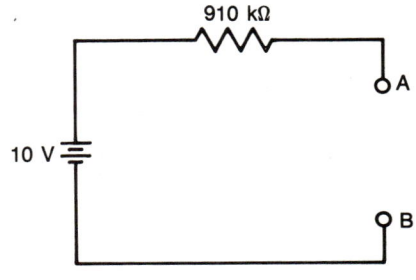

(a) Circuit for Activity 6-2

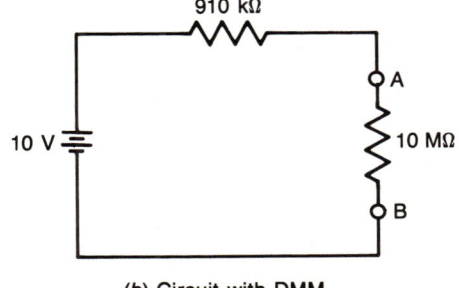

(b) Circuit with DMM

FIGURE 6-8 Activity 6-2.

Procedure

1. Design, construct, and test
   a. A two-resistor circuit in which $I_{R_1} = I_T = 20$ mA and $V_{R_1} = 0.25\, V_T$.
   b. A three-resistor circuit in which $V_{R_1} = 2V_{R_2} = 3V_{R_3}$ and $V_T = 16.5$ V.
2. Appendix 1 provides some guidelines on writing an activity report. You are to produce a report which meets or exceeds those guidelines.

NAME _____   DATE _____

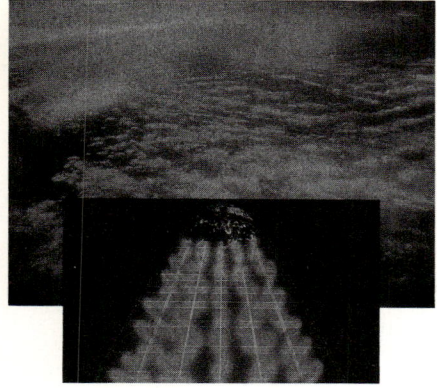

Chapter 7

Parallel Circuits

ACTIVITY 7-1 PARALLEL CIRCUIT CONCEPTS

Introduction

Parallel circuits behave according to Ohm's law and Kirchhoff's current law. These laws are critical for understanding all electrical circuits. This activity investigates the laws for parallel circuits.

Supplies

- (1) Resistor, 220-Ω, ½-W, ±5%
- (1) Resistor, 330-Ω, ½-W, ±5%
- (1) Resistor, 560-Ω, ½-W, ±5%
- (1) Resistor, 1-kΩ, ½-W, ±5%
- (1) Resistor, 10-kΩ, ½-W, ±5%
- (1) Lamp, 14-V, 0.08-A, no. 756, with holder
- (1) DMM or VOM
- (1) Power supply, 0- to 25-V dc

	Calculated	Measured
R_T		
I_T		
I_1		
I_2		
I_3		
P_T		

FIGURE 7-2 Data table.

Procedure

1. Parallel circuit concepts
   a. Connect three resistors in parallel as shown in Fig. 7-1. Do not connect the power supply yet. Calculate and measure the total resistance. Record your results in the table provided in Fig. 7-2. Calculate all of the remaining values shown in the table. Connect the power supply and measure all of the values except power.
   b. Measure the voltage drop across each resistor. Is it the same as the supply voltage? What is constant in a parallel circuit?
   c. Is the total current in the parallel circuit equal to the sum of the branch currents? What is the name given this circuit law?

2. Swamping
   a. Connect two resistors as shown in Fig. 7-3. Measure the resistance of the parallel combination. Suppose the 10-kΩ resistor was burned out (had infinite resistance). Would this fault be detected by measuring the resistors in parallel? Why?

FIGURE 7-1 Parallel circuit.

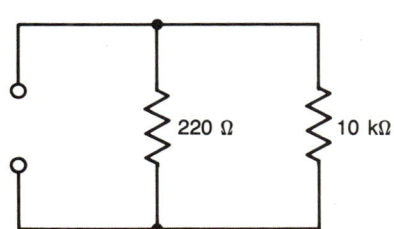

FIGURE 7-3 Swamping effects in parallel circuits.

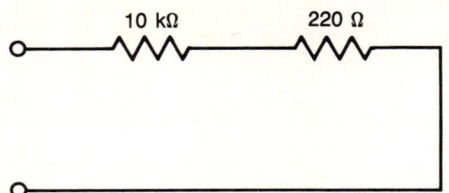

FIGURE 7-4 Swamping effects in series circuits.

b. Remove the 10-kΩ resistor and note that the ohmmeter reading increases just slightly, if at all. Which resistor "swamps" the circuit as far as total resistance is concerned?

c. Suppose the 10-kΩ resistor in Fig. 7-3 is shorted. Can this fault be detected by measuring total resistance? Note: Resistors seldom short but other devices do. How would you proceed to determine which component was shorted in a case such as this?

d. Connect two resistors as shown in Fig. 7-4. Measure their total resistance. Which resistor "swamps" the reading in this case? Would it be

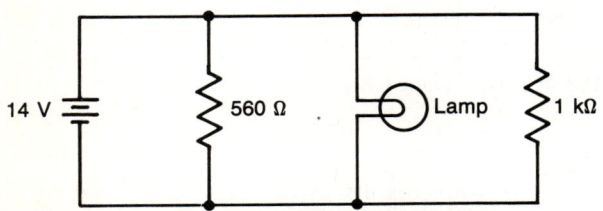

FIGURE 7-5 Simulating parallel circuit faults.

possible to confirm a shorted 220-Ω resistor by measuring total resistance? If the 220-Ω resistor was burned out or very high in value, would this fault be detected by measuring total resistance?

3. Simulating parallel circuit faults
   a. Construct the circuit shown in Fig. 7-5. Note the appearance of the lamp. Simulate an open component by removing the 560-Ω resistor from the circuit. Can you detect any change in the lamp?
   b. Do *not* perform this step if your supply is not current-limited. Short the 1-kΩ resistor. What happens to the lamp output?

ACTIVITY 7-2 THE CURRENT DIVIDER

Introduction

The current divider rule provides a convenient method of analysis for certain situations. This activity uses a dc power supply in its current-limiting mode to act as a constant current source. If you do not have access to a dc power supply with adjustable current limiting, you will not be able to perform this activity. However, you can perform the requested calculation and answer the questions.

Supplies

(1) Resistor, 100-Ω, ½-W, ±5%
(1) Resistor, 33-Ω, ½-W, ±5%
(1) DMM or VOM
(1) Power supply, 0- to 25-V dc (with adjustable current limiting)

Procedure

1. Applying the current divider rule
   a. Make *sure* that your supply has adjustable current limiting.
   b. Set the power supply to 10 V.
   c. Set your multimeter to a dc current range greater than 50 mA.
   d. Turn the power supply current limit control all the way down.
   e. Connect the meter to the supply. Adjust the current limit control on the supply for a meter reading of 50 mA.
   f. Connect the circuit shown in Fig. 7-6(a). Note: The symbol for a current source is a circle with an arrow in it. This represents the power supply operating in current limiting. The current flow in the resistor should be 50 mA.
   g. Look at the circuit shown in Fig. 7-6(b). Use the current divider formula to predict the current flow in the 100-Ω resistor. Build the circuit. The measured value should be reasonably close to the predicted value.
   h. Could you have predicted the meter reading in Fig. 7-6(b) without using the current divider rule? How?

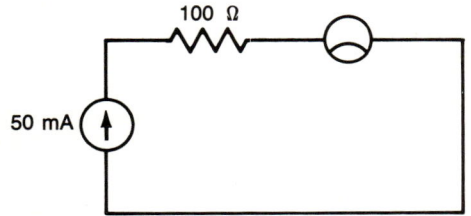

(a) Circuit with one path

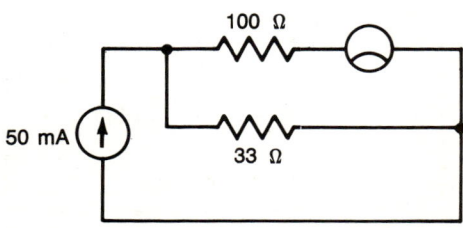

(b) Circuit with two paths

FIGURE 7-6 Activity 7-2.

ACTIVITY 7-3 DESIGNING PARALLEL CIRCUITS

Introduction

Previous activities for parallel circuits specified the circuit diagrams, component values, and source conditions to be used, as well as a step-by-step procedure to be followed. In this activity, only the circuit characteristics are specified. You must design the circuit, select the components, and develop the procedure needed to verify your design experimentally.

Supplies

(1) Power supply, 0- to 25-V dc

(1) DMM or VOM

Miscellaneous components selected from the Materials List in this manual

Procedure

1. Design, construct, and test
   a. A two-resistor circuit in which $V_{R_1} = 12$ V, $I_{R_2} = 10$ mA, and $3I_{R_1} = 2I_{R_2}$.
   b. A two-resistor circuit in which $P_{R_1} = 0.1$ W $= 3P_{R_2}$.
2. Prepare a report that meets or exceeds the guidelines given in Appendix 1.

NAME _____ DATE _____

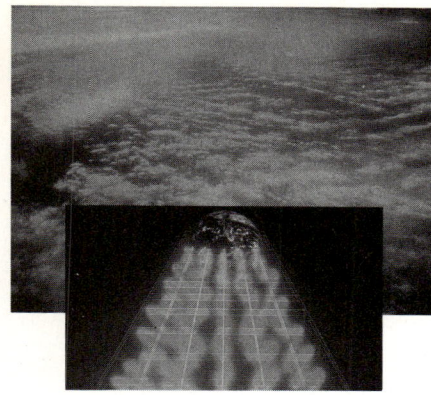

Chapter 8

Series-Parallel Circuits

ACTIVITY 8-1   SERIES-PARALLEL CIRCUIT CONCEPTS

Introduction

Series-parallel circuits behave according to Ohm's law and Kirchhoff's laws. Most practical electrical and electronic circuits have both series and parallel sections. Technicians must be able to analyze, construct, measure, and troubleshoot series-parallel circuits.

Supplies

(2)   Resistors, 220-Ω, ½-W, ±5%
(1)   Resistor, 560-Ω, ½-W, ±5%
(1)   Resistor, 820-Ω, ½-W, ±5%
(2)   Resistors, 1-kΩ, ½-W, ±5%
(1)   DMM or VOM
(1)   Power supply, 0- to 25-V dc

Procedure

1. Series-parallel circuit concepts
    a. Construct the circuit shown in Fig. 8-1. Do not connect the power supply yet. Calculate and measure total resistance. Record your results in the table provided in Fig. 8-2. Calculate all of the remaining values shown in the table. Connect the power supply and measure all of the values.
    b. What is constant in the series section of the circuit? The parallel section?
2. Faults in series-parallel circuits
    a. Measure the voltage across R_5. Short R_6. What happens to the measured voltage? Why?
    b. Remove the short. Measure the drop across R_5. Open R_1. What happens to the measured voltage? Why?

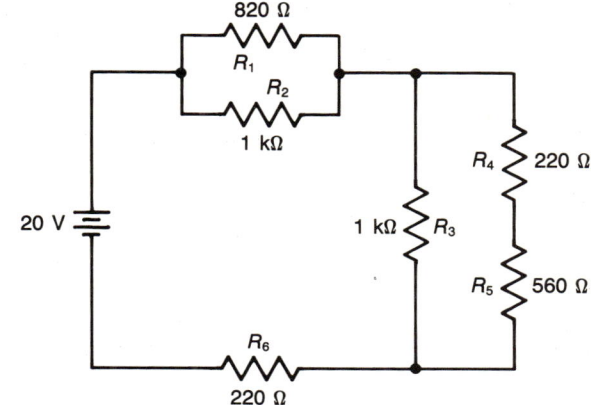

FIGURE 8-1   Series-parallel circuit.

	Calculated	Measured
R_T		
I_T		
V_{R_1}		
V_{R_2}		
V_{R_3}		
V_{R_4}		
V_{R_5}		
V_{R_6}		
I_{R_1}		
I_{R_2}		
I_{R_3}		
I_{R_4}		
I_{R_5}		
I_{R_6}		

FIGURE 8-2   Data table

Copyright © 1993 by the Glencoe Division of Macmillan/McGraw-Hill School Publishing Company. All rights reserved.

ACTIVITY 8-2   VOLTAGE DIVIDERS

Introduction

Loaded voltage dividers are series-parallel circuits. They are often designed with a bleeder resistor to improve their voltage regulation. Voltage regulation is the ability of a supply or a circuit to maintain a constant output voltage. This activity investigates loaded voltage dividers with and without bleeder current.

Supplies

- (1) Resistor, 33-Ω, $\frac{1}{2}$-W, ±5%
- (1) Resistor, 43-Ω, $\frac{1}{2}$-W, ±5%
- (2) Resistors, 220-Ω, $\frac{1}{2}$-W, ±5%
- (1) Resistor, 300-Ω, $\frac{1}{2}$-W, ±5%
- (1) Resistor, 330-Ω, $\frac{1}{2}$-W, ±5%
- (1) Resistor, 560-Ω, $\frac{1}{2}$-W, ±5%
- (1) Resistor, 1-kΩ, $\frac{1}{2}$-W, ±5%
- (1) DMM or VOM
- (1) Power supply, 0- to 25-V dc

Procedure

1. **A voltage divider with bleeder current**
   a. Figure 8-3 shows a voltage divider schematic. Calculate V_A and V_B for this circuit. Build the circuit and measure these voltages to verify your calculations.
   b. The current sum for the 330- and 560-Ω resistors makes up the total load current in Fig. 8-3. The current in the 1-kΩ resistor is the bleeder current. Calculate the percentage of bleeder current for this circuit.

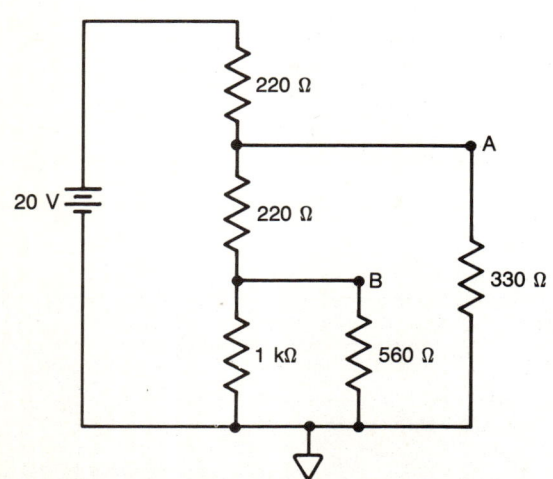

FIGURE 8-3   Voltage divider.

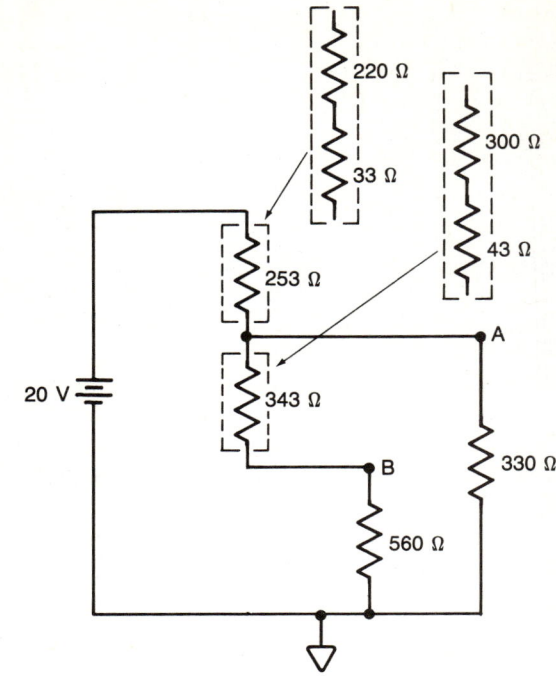

FIGURE 8-4   Voltage divider with no bleeder current.

   c. Remove the 560-Ω load resistor from the circuit. Remeasure V_B. Calculate the percentage of voltage regulation by dividing the change in V_B by its original value.
   d. With the 560-Ω load resistor still out of the circuit, remeasure V_A. Has it changed from its original value? Is there interaction among the loads in voltage dividers of this type?

2. **A voltage divider without bleeder current**
   a. Figure 8-4 shows another voltage divider circuit. Calculate V_A and V_B for this circuit. Build the circuit and measure these voltages to verify your calculations. Note: 253 Ω and 343 Ω are nonstandard resistor values. They are obtained by series-connecting standard values.
   b. What is the percentage of bleeder current in Fig. 8-4?
   c. Remove the 560-Ω resistor from the circuit. Remeasure V_B. Calculate the percentage of voltage regulation for this circuit.
   d. An ideal voltage source shows no change when its load changes. Its voltage regulation is 0. Which of the two voltage-divider circuits exhibits the best voltage regulation?

ACTIVITY 8-3   THE WHEATSTONE BRIDGE CIRCUIT

Introduction

The Wheatstone bridge is very popular in instrumentation applications. A sensor can be connected in one arm of the bridge. When the sensor changes, the bridge output voltage changes. This output signal is proportional to the quantity being sensed. A bridge may be nulled (set for zero output voltage). This activity investigates this procedure.

Supplies

(1) Resistor, 27-Ω, $\frac{1}{2}$-W, ±5%
(1) Resistor, 33-Ω, $\frac{1}{2}$-W, ±5%
(1) Resistor, 560-Ω, $\frac{1}{2}$-W, ±5%
(1) Resistor, 10-kΩ, $\frac{1}{2}$-W, ±5%
(1) Potentiometer, 1-kΩ, $\frac{1}{2}$-W
(1) DMM or VOM
(1) Power supply, 0- to 25-V dc

Procedure

1. Nulling a bridge
   a. Build the circuit shown in Fig. 8-5. Calculate the value for the potentiometer that will null the bridge. Null the bridge by adjusting the potentiometer for a voltage reading of zero. Remove the potentiometer from the bridge circuit and be careful not to disturb its setting. Measure its

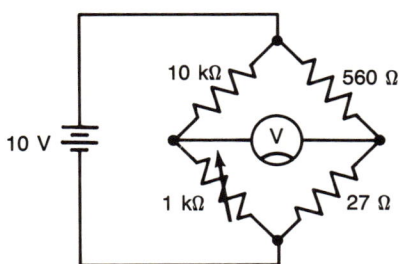

FIGURE 8-5 Wheatstone bridge.

resistance with your multimeter. Your measured value should agree with your calculated value.

b. Return the potentiometer to the bridge circuit. Replace the 27-Ω resistor with a 33-Ω resistor. Calculate the value required for the potentiometer to null the bridge. Null the bridge with the potentiometer. Carefully remove it from the circuit and measure its value to verify your calculations.

c. The change in one leg of the bridge was 6 Ω (33 − 27). What change was required in the adjustable leg of the bridge to reestablish the null condition?

d. Bridge circuits can be used to measure very small values of resistance and very small changes in resistance. Why?

ACTIVITY 8-4 DESIGNING A SERIES-PARALLEL CIRCUIT

Introduction

In this activity you are to select the components, circuit configuration, and procedures needed to produce the specified results. Rather than responding to specific questions about a given concept, you will write a detailed report to show that you understand the concepts involved in the activity.

Supplies

(1) Power supply, 0- to 25-V dc
(1) DMM or VOM

Miscellaneous resistors selected from the Materials List in this manual

Procedure

1. Design, construct, and test a series-parallel resistor circuit in which $I_{R_2} = I_{R_3} = 0.5I_{R_1}$ and $5V_{R_1} = 3V_{R_2}$.
2. Write an activities report which meets or exceeds the requirements listed in Appendix 1.

NAME _____   DATE _____

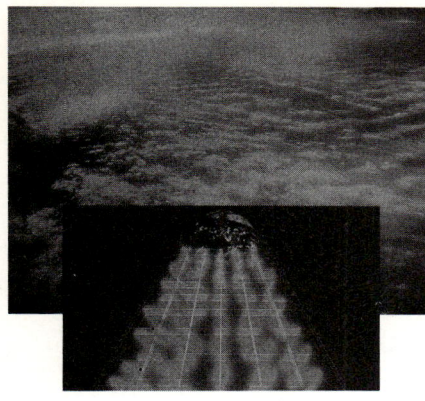

Chapter 9

Network Analysis

ACTIVITY 9-1   CONSTANT CURRENT SOURCE

Introduction

This activity will acquaint you with the primary characteristic of a constant current source: an output current that does not vary with changing loads. It will also demonstrate one of the limitations of constant current sources. You will build a practical constant current source using a transistor, a zener diode, and two resistors.

Supplies

(1) Power supply, 0- to 25-V dc, 0.5-A capacity, floating output
(1) PNP transistor, 2N4126 or equivalent
(1) Zener diode, 5.6-V, 1-W, ±5%, 1N4733B or equivalent
(1) Resistor, 100-Ω, ½-W, ±5%
(1) Resistor, 560-Ω, ½-W, ±5%
(2) Resistors, 1-kΩ, ½-W, ±5%
(1) Resistor, 1.5-kΩ, ½-W, ±5%
(1) Resistor, 2.2-kΩ, ½-W, ±5%
(1) Resistor, 3.3-kΩ, ½-W, ±5%
(1) Resistor, 5.1-kΩ, ½-W, ±5%
(1) Resistor, 10-kΩ, ½-W, ±5%
(1) VOM
(1) DMM

Procedure

1. Build the circuit shown in Fig. 9-1. Both Z_1, the zener diode, and Q_1, the transistor, are polarized devices so you must identify the leads before you place the devices in the circuit. Refer to Fig. 9-2 to identify the leads. Make sure you have identified the leads properly; if in doubt, ask the instructor

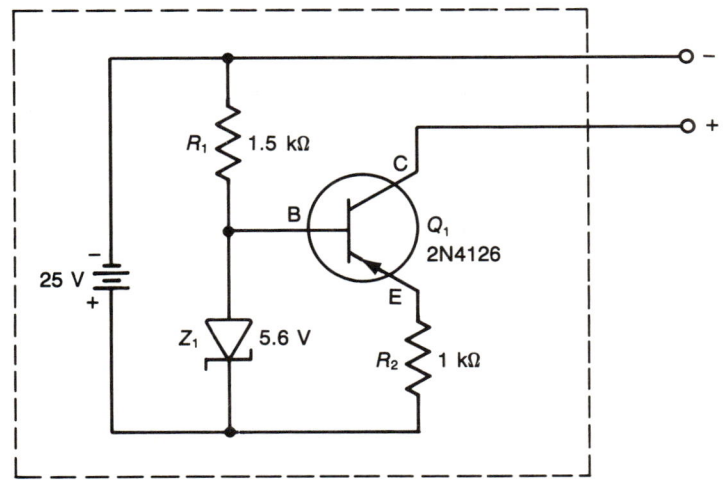

(a) Schematic diagram

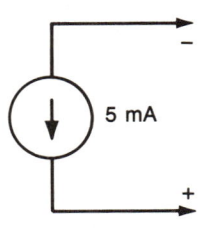

(b) Equivalent circuit

FIGURE 9-1 Constant current source for Activity 9-1. All circuitry inside the broken lines, including the voltage source, comprises the current source.

Copyright © 1993 by the Glencoe Division of Macmillan/McGraw-Hill School Publishing Company. All rights reserved.

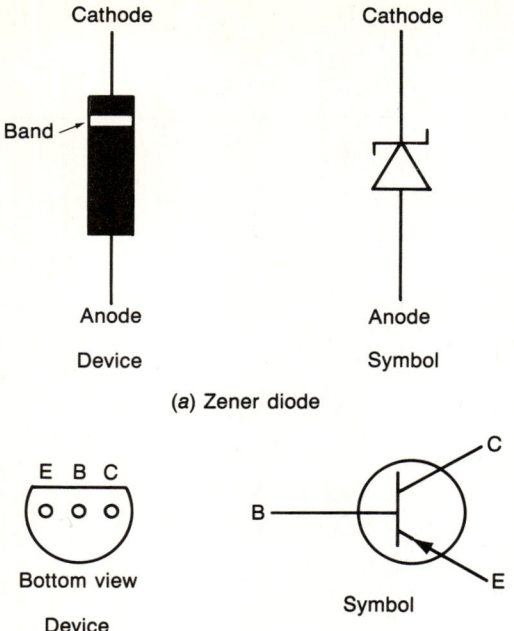

FIGURE 9-2 Lead identification for the zener diode and PNP transistor used in Activity 9-1.

for help. Use the DMM to adjust the power supply for 25 V.

2. Measure and record in Table 9-1 the short-circuit current that the circuit will supply by connecting the VOM across the output terminals.
3. Now connect the various load resistors specified in Table 9-1 and measure the current through them with the VOM and the voltage across them with a DMM. Record the readings in the table.
4. For load resistances lower than 3.5 kΩ, does the circuit exhibit constant current or constant voltage characteristics?
5. When the load resistance becomes too large, the source loses its constant current characteristics and the load current decreases. Did your circuit ever reach this limit? What was the load voltage at this limit? What value of load resistor would just reach this upper limit and still maintain a constant current output? (Hint: use Ohm's law to calculate R.)
6. Remove the last load resistor from the constant current source but *do not* disassemble the current source. It will be used in Activities 9-2 and 9-6.

ACTIVITY 9-2 DUALS

Introduction

This activity will provide experience in calculating the *dual* of a constant current source. It will allow you to verify your calculations by measuring currents and voltages for various loads connected to dual constant voltage and constant current sources.

Supplies

(1) Power supply, 0- to 25-V dc, 0.5-A capacity, floating output
(1) PNP transistor, 2N4126 or equivalent
(1) Zener diode, 5.6-V, 1-W, ±5%, 1N4733B or equivalent
(3) Resistors, 1-kΩ, ½-W, ±5%
(1) Resistor, 1.5-kΩ, ½-W, ±5%
(1) Resistor, 2.2-kΩ, ½-W, ±5%
(1) VOM
(1) DMM

Procedure

1. Figure 9-3(b) shows a 5-mA source in shunt with a 1-kΩ resistor. On a separate sheet of paper, draw a schematic of the *dual* voltage source for the current source of Fig. 9-3(b). Now calculate the voltage and series resistance of the dual voltage source.
2. Build the circuit shown in Fig. 9-3(a). Note: This is the current source of Activity 9-1 with a shunt resistance of 1 kΩ placed across the output terminals. Measure and record the open-circuit voltage of this source by connecting a DMM across the output terminals. Now measure and record the short-circuit current of this source by connecting a VOM set to the 10-mA current range across the output leads.
3. Connect a 1-kΩ resistor to the output terminals of the current source; then measure and record the current through the resistor. Also measure and record the voltage across the resistor.

TABLE 9-1

Load Resistance, Ω	Load Current	Load Voltage
0 (short circuit)		
100		
560		
1000		
2200		
3300		
5100		
10,000		

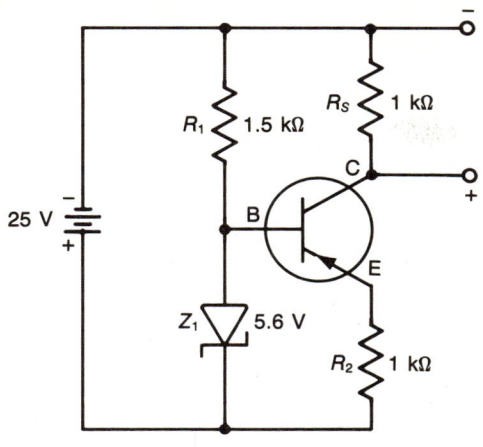
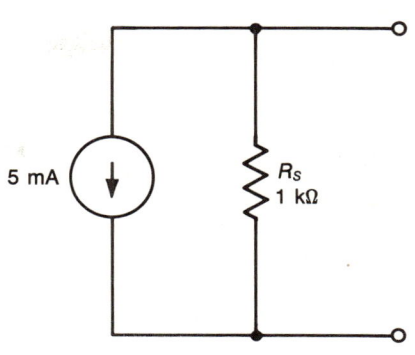

(a) Schematic diagram (b) Equivalent circuit

FIGURE 9-3 Constant current source with 1-kΩ shunt resistor.

4. Replace the 1-kΩ resistor with a 2.2-kΩ resistor and again measure and record the current through and the voltage across the resistor.

5. Build the dual voltage source that you designed in step 1. Measure and record the open-circuit voltage of this source. Now measure and record the short-circuit current of this source. If you calculated the dual correctly, the measurements should be approximately the same as those you obtained in step 2.

6. Connect a 1-kΩ resistor to the output terminals of the dual voltage source; then measure and record the current through the resistor. Also measure and record the voltage across the resistor.

7. Replace the 1-kΩ resistor with a 2.2-kΩ resistor and again measure and record the current through and the voltage across the resistor. Do the values measured for the dual voltage source match the values measured for the constant current source? If they match within ±10%, then the two circuits are duals and your calculations are correct.

ACTIVITY 9-3 MESH ANALYSIS

Introduction

This activity will provide experience in calculating and measuring currents and voltages in a complex circuit. The measured values will verify whether or not the calculations are correct, within the tolerances of the components and equipment.

Supplies

(1) Power supply, 0- to 25-V dc, 0.5-A capacity, floating output

(4) Alkaline D cells with holders or a second power supply
(2) Resistors, 1-kΩ, ½-W, ±5%
(1) Resistor, 1.2-kΩ, ½-W, ±5%
(1) Resistor, 2.2-kΩ, ½-W, ±5%
(1) VOM
(1) DMM

Procedure

1. Apply mesh analysis to solve the network shown in Fig. 9-4. Calculate and record the following currents: I_{R_1}, I_{R_2}, I_{R_3}, and I_{R_4}. Calculate and record the following voltages: V_{R_1}, V_{R_2}, V_{R_3}, and V_{R_4}. What is the polarity of point A with respect to point B?

2. Construct the circuit shown in Fig. 9-4. Use a 0- to 25-V power supply for the 10-V volt source and use four alkaline D cells connected in series for the 6-V source.

3. Using a DMM, measure and record the following voltages: B_1, B_2, V_{R_1}, V_{R_2}, V_{R_3}, and V_{R_4}. Compare the measured voltages with the voltages calculated in step 1. They should agree rather closely.

4. Using a VOM, measure and record the currents through the following resistors: R_1, R_2, R_3, and R_4. Compare the measured currents with the currents calculated in step 1. They should also agree rather closely. If any discrepancy occurs, the problem may be an incorrect mesh analysis solution or an incorrect meter connection. Remember: to measure the current you must break the circuit and insert the meter in series with the resistor!

5. Is the polarity of V_{R_3} the same as you predicted in step 1? Does Kirchhoff's voltage law apply to the loop containing R_3, R_4 and B_2?

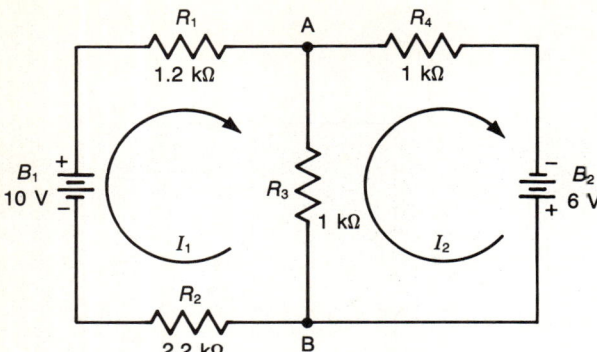

FIGURE 9-4 Complex circuit for Activity 9-3.

ACTIVITY 9-4   SUPERPOSITION THEOREM

Introduction

This activity will provide additional experience in calculating and measuring currents and voltages in a complex circuit that contains two power supplies. In this exercise you will experimentally verify the techniques used in applying the superposition theorem.

Supplies

(1) Power supply, 0- to 25-V dc, 0.5-A capacity, floating output
(4) Alkaline D cells with holders or a second power supply
(1) Resistor, 1-kΩ, ½-W, ±5%
(1) Resistor, 1.2-kΩ, ½-W, ±5%
(1) Resistor, 1.5-kΩ, ½-W, ±5%
(1) VOM
(1) DMM

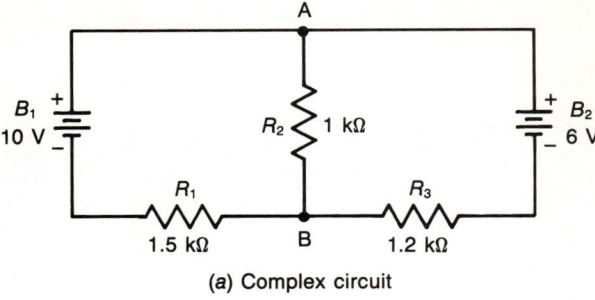

(a) Complex circuit

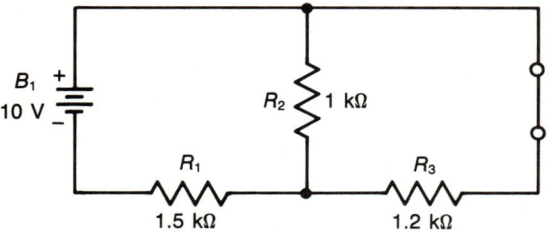

(b) Series-parallel circuit with source B_2 removed and replaced with a short circuit

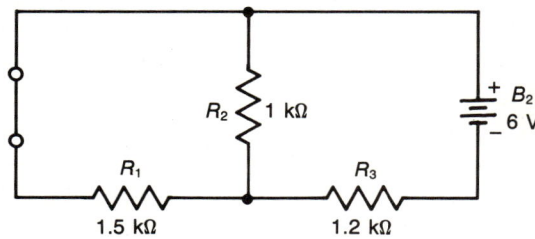

(c) Series-parallel circuit with source B_1 removed and replaced with a short circuit

FIGURE 9-5 Circuits for Activity 9-4.

Procedure

1. To solve the complex circuit shown in Fig. 9-5(a) using superposition, the first step is to remove one of the voltage sources and replace it with a short circuit. Figure 9-5(b) shows the circuit with source B_2 removed and replaced with a jumper. Solve this series-parallel circuit for all currents and record the results on a separate sheet of paper.

2. Build the series-parallel circuit shown in Fig. 9-5(b) and verify the magnitudes and directions of the currents that you solved for in step 1. The currents should agree within ±10% of those calculated.

3. Figure 9-5(c) shows the circuit with source B_2 returned to the circuit and source B_1 removed and replaced with a jumper. Solve this series-parallel circuit for all currents and record the results on a separate sheet of paper.

4. Build the series-parallel circuit shown in Fig. 9-5(c) and verify the magnitudes and directions of the currents that you solved for in step 3. The currents should agree within ±10% of those calculated.

5. On a separate sheet of paper, copy the complex circuit of Fig. 9-5(a) and algebraically superimpose the currents found in steps 1 and 3. Verify the results by building the complex circuit using two power supplies; use the 0- to 25-V supply for the 10-V battery and use four alkaline D cells in series for the 6-V battery. Measure and record the direction and magnitudes of the currents through R_1, R_2, and R_3. Do the measured currents agree with the calculated currents? What is the polarity of point A with respect to point B?

6. List the primary advantage of using superposition over mesh analysis to solve a complex circuit.

NAME _____ DATE _____

ACTIVITY 9-5 THEVENIN'S THEOREM

Introduction

This activity will provide experience in converting complex circuits into equivalent circuits.

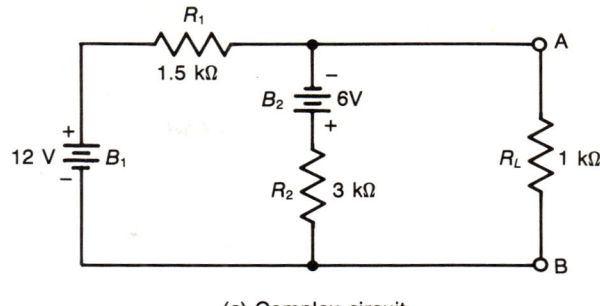

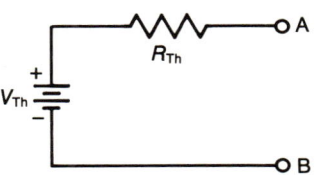

FIGURE 9-6 Circuits for Activity 9-5.

Supplies

(1) Power supply, 0 to 25-V dc, 0.5-A capacity, floating output
(4) Alkaline D cells with holders or a second power supply
(2) Resistors, 1-kΩ, $\frac{1}{2}$-W, $\pm 5\%$
(1) Resistor, 1.5-kΩ, $\frac{1}{2}$-W, $\pm 5\%$
(1) Resistor, 2.2-kΩ, $\frac{1}{2}$-W, $\pm 5\%$
(1) Resistor, 3-kΩ, $\frac{1}{2}$-W, $\pm 5\%$
(1) Resistor, 3.3-kΩ, $\frac{1}{2}$-W, $\pm 5\%$
(1) VOM
(1) DMM

Procedure

1. Construct the complex circuit drawn in Fig. 9-6(a). Temporarily remove the load resistor (R_L) located between points A and B. Use the DMM to measure the open-circuit voltage and record the value in the appropriate box in Table 9-2. Use the VOM to measure the short-circuit current of the complex circuit and record it in the table also.

2. Reconnect the 1-kΩ load resistor between points A and B of the complex circuit and measure and record V_{R_L} and I_{R_L}. Now change R_L to 2.2 kΩ and again measure and record V_{R_L} and I_{R_L}. Finally, change R_L to 3.3 kΩ and measure and record V_{R_L} and I_{R_L}. *Do not disassemble this circuit until you are instructed to do so.*

3. Using R_L as the load resistor, find the Thevenin equivalent circuit for the schematic drawn in Fig. 9-6(a). Record your calculated values of V_{Th} and R_{Th}. V_{Th} should equal the open-circuit voltage that you measured in step 1. R_{Th} can be measured in the following manner: (1) remove supplies B_1 and B_2 from the circuit of Fig. 9-6(a), (2) replace them with jumper wires, (3) remove the load resistor from the circuit, and (4) measure the resistance between points A and B with the ohmmeter function of the DMM.

4. Using the Thevenin equivalent circuit developed in step 3, calculate all voltages and currents needed to complete row 2 of Table 9-2.

5. Disassemble the circuit of Fig. 9-6(a) and build the circuit drawn in Fig. 9-6(b) using the values for V_{Th} and R_{Th} that you calculated in step 3. With no load resistor connected, measure and record (in Table 9-2) the open-circuit voltage and the short-

TABLE 9-2

Type of Circuit	Open-Circuit Voltage	Short-Circuit Current	Value of R_L					
			1000 Ω		2200 Ω		3300 Ω	
			V_{R_L}	I_{R_L}	V_{R_L}	I_{R_L}	V_{R_L}	I_{R_L}
Complex Fig. 9-6(a) (measured)								
Thevenin Fig. 9-6(b) (calculated)								
Thevenin Fig. 9-6(b) (measured)								

circuit current for the equivalent circuit. Does the open-circuit voltage and short-circuit current for Fig. 9-6(b) agree, within ±10%, with those measured for Fig. 9-6(a)?

6. Should a 1-kΩ load across terminals A and B of Fig. 9-6(b) receive the same voltage and current as the 1-kΩ load (R_L) in Fig. 9-6(a)?

7. Connect the three different values of R_L to the circuit shown in Fig. 9-6(b) and make the measurements necessary to complete Table 9-2. Do the data in rows 1, 2, and 3 of Table 9-2 support Thevenin's theorem?

8. Since Thevenin equivalent circuits and Norton equivalent circuits are *duals*, calculate I_N and R_N for the circuit shown in Fig. 9-6(b) and draw the Norton equivalent circuit.

ACTIVITY 9-6   NORTON'S THEOREM

Introduction

This activity will provide additional experience in converting complex circuits into equivalent circuits. Within the tolerances of the components and equipment, measured values will verify the necessary calculations.

Supplies

(1) Power supply, 0- to 25-V dc, 0.5-A capacity, floating output
(4) Alkaline D cells with holders or a second power supply
(1) PNP transistor, 2N4126 or equivalent
(1) Zener diode, 5.6-V, 1-W, ±5%, 1N4733B or equivalent
(1) Resistor, 560-Ω, ½-W, ±5%
(2) Resistors, 1-kΩ, ½-W, ±5%
(1) Resistor, 1.5-kΩ, ½-W, ±5%
(1) Resistor, 1.8-kΩ, ½-W, ±5%
(2) Resistors, 2.2-kΩ, ½-W, ±5%
(1) Resistor, 10-kΩ, ½-W, ±5%
(1) VOM
(1) DMM

Procedure

1. Construct the complex circuit drawn in Fig. 9-7(a). Temporarily remove the load resistor (R_L) located between points A and B. Use the DMM to measure the open-circuit voltage and record the value in the appropriate box in Table 9-3. Use the VOM to measure the short-circuit current of the complex circuit and record it in the table also.

2. Reconnect the 560-Ω load resistor between points A and B of the complex circuit and measure and record V_{R_L} and I_{R_L}. Now change R_L to 1 kΩ and again measure and record V_{R_L} and I_{R_L}. Finally, change R_L to 2.2 kΩ and measure and record V_{R_L} and I_{R_L}. *Do not disassemble this circuit until you are instructed to do so.*

3. Using R_L as the load resistor, find the Norton equivalent circuit for the schematic drawn in Fig. 9-7(a). Record your calculated values of I_N and R_N. I_N should equal the short-circuit current that you measured in step 1. Resistance R_N can be measured in the following manner: (1) remove supplies B_1 and B_2 from the circuit of Fig. 9-7(a), (2) re-

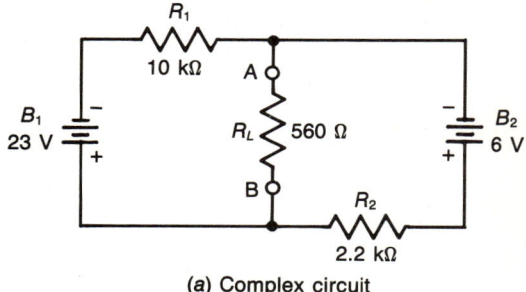

(a) Complex circuit

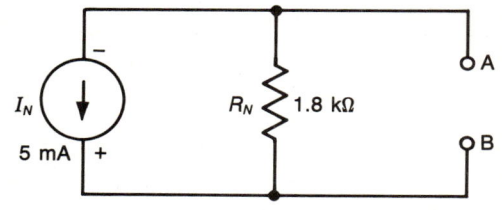

(b) Norton equivalent circuit

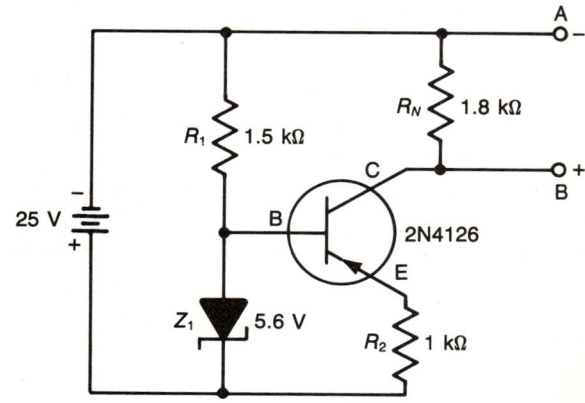

(c) Practical Norton equivalent circuit

FIGURE 9-7 Circuits for Activity 9-6.

TABLE 9-3

Type of Circuit	Open-Circuit Voltage	Short-Circuit Current	Value of R_L					
			560 Ω		1000 Ω		2200 Ω	
			V_{R_L}	I_{R_L}	V_{R_L}	I_{R_L}	V_{R_L}	I_{R_L}
Complex Fig. 9-7(a) (measured)								
Norton Fig. 9-7(b) (calculated)								
Norton Fig. 9-7(c) (measured)								

place them with jumper wires, (3) remove the load resistor from the circuit, and (4) measure the resistance between points A and B with the ohmmeter function of the DMM.

4. The Norton equivalent circuit you developed in step 3 should look like the one detailed in Fig. 9-7(b). If it doesn't, try again! Using the Norton equivalent circuit of Fig. 9-7(b), calculate all voltages and currents needed to complete row 2 of Table 9-3.

5. Figure 9-7(c) shows a practical circuit diagram for the Norton equivalent circuit of Fig. 9-7(b). It is composed of the 5-mA constant current source used in Activity 9-1 and an added 1.8-kΩ (R_N) of internal parallel resistance. Thus, Fig. 9-7(b) and (c) are electrically equivalent to Fig. 9-7(a) when R_L is removed from the circuit.

6. Disassemble the circuit of Fig. 9-7(a) and build the circuit shown in Fig. 9-7(c). With no load resistor connected, measure and record (in Table 9-3) the open-circuit voltage and the short-circuit current for the practical Norton equivalent circuit. Does the open-circuit voltage and short current for Fig. 9-7(c) agree, within ±10%, with those measured for Fig. 9-7(a)?

7. Should a 560-Ω load across terminals A and B of Fig. 9-7(c) receive the same voltage and current as the 560-Ω load (R_L) in Fig. 9-7(a)?

8. Connect the three different values of R_L to the circuit shown in Fig. 9-7(c) and make the measurements necessary to complete Table 9-3. Do the data in rows 1, 2, and 3 of Table 9-3 support Norton's theorem?

9. Since Norton equivalent circuits and Thevenin equivalent circuits are *duals*, calculate V_{Th} and R_{Th} for the circuit shown in Fig. 9-7(b) and draw the Thevenin equivalent circuit.

ACTIVITY 9-7 NETWORK TRANSFORMS

Introduction

This activity will provide you with experience in calculating delta-wye transforms. You will also verify experimentally if the calculations are correct.

Supplies

(2) Resistors, 10-Ω, ½-W, ±5%
(1) Resistor, 100-Ω, ½-W, ±5%
(1) Resistor, 300-Ω, ½-W, ±5%
(1) Resistor, 560-Ω, ½-W, ±5%
(1) Resistor, 820-Ω, ½-W, ±5%
(3) Resistors, 1-kΩ, ½-W, ±5%
(1) Resistor, 1.2-kΩ, ½-W, ±5%
(1) Resistor, 1.5-kΩ, ½-W, ±5%
(3) Resistors, 3-kΩ, ½-W, ±5%
(1) Resistor, 3.3-kΩ, ½-W, ±5%
(1) DMM

Procedure

1. Examine the delta network shown in Fig. 9-8(a). Copy the wye network shown in Fig. 9-8(b) onto a separate sheet of paper. Apply the delta to wye transformation equations and solve for the values of the following resistors: R_4, R_5, and R_6.

2. Build the delta network shown in Fig. 9-8(a) and the wye network shown in Fig. 9-8(b). Use the calculated values of resistors for the wye network. Using a DMM, measure and record the resistance between points A and B. Now measure and re-

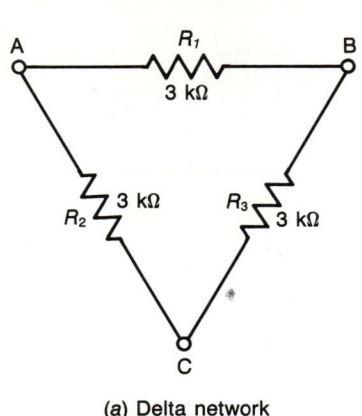

(a) Delta network

(b) Wye network

FIGURE 9-8 Circuits for Activity 9-7. Delta to wye transforms.

cord the resistance between points X and Y. Are the values nearly the same?

3. Measure and record the resistance between points B and C; measure and record the resistance between points Y and Z. Are the measured values nearly the same?

4. Measure and record the resistance between points C and A; measure and record the resistance between points Z and X. Are these measured values nearly the same?

5. Is the wye network you designed equivalent to the original delta network?

6. Examine the wye network shown in Fig. 9-9(a). Copy the delta network shown in Fig. 9-9(b) onto a separate sheet of paper. Apply the wye to delta transformation equations and solve for the values of the following resistors: R_{10}, R_{11}, and R_{12}.

7. Build the wye network shown in Fig. 9-9(a); use a 560-Ω resistor in series with a 100-Ω resistor to form resistor R_8. To form resistor R_9, place two parallel-connected 10-Ω resistors in series with an 820-Ω resistor. Build the delta network shown in Fig. 9-9(b); if you did your calculations correctly, the values of resistors 10, 11, and 12 should have come out to standard nominal resistance values.

8. Measure and record the resistance between points D and E; measure and record the resistance between points R and S. Are the measured values nearly the same? If they aren't, either your calculations are incorrect, the circuit is wired incorrectly, or the resistors are out of tolerance. Please make sure the resistors are in the proper locations.

9. Measure and record the resistance between points E and F; measure and record the resistance between points S and T. Are the measured values nearly the same?

10. Measure and record the resistance between points F and D; measure and record the resist-

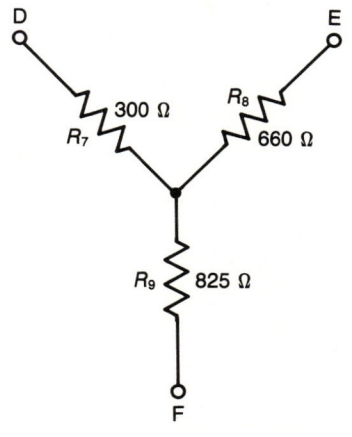

(a) Wye network

(b) Delta network

FIGURE 9-9 Circuits for Activity 9-7. Wye to delta transforms.

ance between points T and R. Are the measured values nearly the same? Is the delta network you designed equivalent to the original wye network?

ACTIVITY 9-8 ANALYZING AN UNBALANCED BRIDGE CIRCUIT

Introduction

This activity will provide experience in calculating and measuring currents and voltages in a complex circuit. You will solve the unbalanced bridge circuit using delta-wye transforms to determine the total resistance in the circuit and then apply the current divider formula, Ohm's law, and Kirchhoff's laws to solve for all unknown quantities. Within the tolerances of the equipment and components, the measured values will verify whether or not the calculations are correct.

Supplies

(1) Power supply, 0- to 25-V dc, 0.5-A capacity, floating output
(1) Resistor, 1-kΩ, $\frac{1}{2}$-W, $\pm5\%$
(1) Resistor, 1.2-kΩ, $\frac{1}{2}$-W, $\pm5\%$
(1) Resistor, 1.5-kΩ, $\frac{1}{2}$-W, $\pm5\%$
(1) Resistor, 3-kΩ, $\frac{1}{2}$-W, $\pm5\%$
(1) Resistor, 3.3-kΩ, $\frac{1}{2}$-W, $\pm5\%$
(1) VOM
(1) DMM

Procedure

1. Examine the circuit shown in Fig. 9-10(a). It is an unbalanced bridge circuit which is an example of an irreducible circuit. By applying the delta to wye transformation equations to the top delta section of the bridge, it is possible to obtain an equivalent circuit that is reducible. Figure 9-10(b) shows the top delta section removed from the bridge and the equivalent wye network. This is the same delta and wye from the second half of Fig. 9-9. You already calculated the equivalent resistors in the previous lab activity so the values are listed on the schematic.

2. Figure 9-10(c) shows the bridge circuit with the top delta section replaced by the equivalent wye. This circuit is reducible. On a separate sheet of paper copy the schematics shown in Fig. 9-10(a) and (c). Using the circuit shown in Fig. 9-10(c), calculate

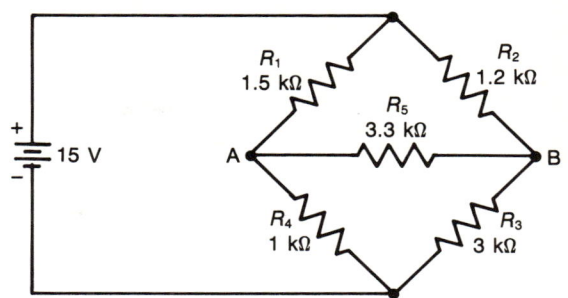

(a) An unbalanced bridge circuit

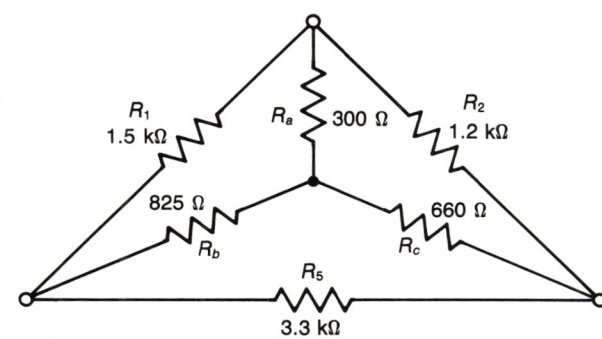

(b) The top delta section and its equivalent wye

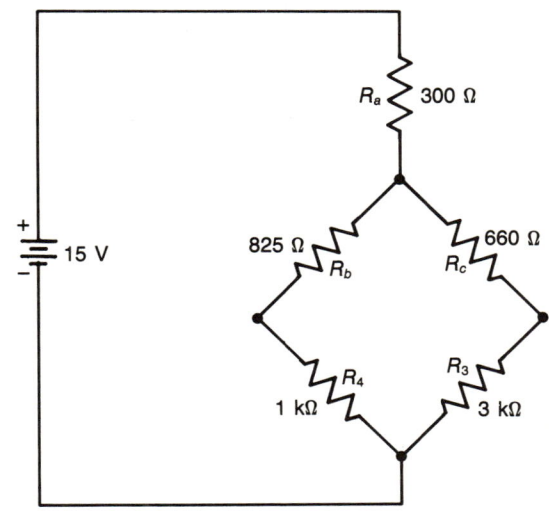

(c) The bridge circuit with the top delta section replaced with its equivalent wye

FIGURE 9-10 Circuits for Activity 9-8.

the total resistance and record it in Table 9-4. Solve for the total current in the circuit and record it. Use the current divider formula to find the currents through the branches containing R_3 and R_4; record them in Table 9-4.

3. Transfer the calculated values of I_T, I_{R_3}, and I_{R_4} to your copy of Fig. 9-10(a). Use Ohm's law and

TABLE 9-4

	Calculated	Measured
R_T		
I_T		
I_{R_1}		
V_{R_1}		
I_{R_2}		
V_{R_2}		
I_{R_3}		
V_{R_3}		
I_{R_4}		
V_{R_4}		
I_{R_5}		
V_{R_5}		

Kirchhoff's voltage and current laws to solve for the remaining voltages and currents.

4. Build the circuit shown in Fig. 9-10(a). Temporarily disconnect the power supply from the circuit and, using the ohmmeter function of the DMM, measure and record the total resistance of the bridge circuit. Reconnect the power supply to the circuit and adjust it for 15 V. Measure and record all voltages and currents necessary to complete Table 9-4.

5. Do the calculated and measured values in Table 9-4 agree rather closely? If these values differ by more than ±10%, check your measurement techniques and your calculations for errors.

6. What additional methods of network analysis could be used to solve the unbalanced bridge circuit?

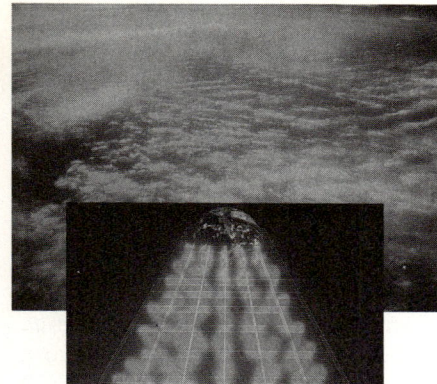

Chapter 10

Magnetism

ACTIVITY 10-1 PERMANENT MAGNETS

Introduction

This activity deals with some of the qualitative characteristics of magnets and materials. Quantitative experiments require materials and equipment that are not normally available in school laboratories. Some of the operations will be performed with a magnetic compass. Note: A magnetic compass can be damaged by a strong external field. Do not bring a powerful magnet close to a compass.

Supplies

(2) Bar magnets
(1) Magnetic compass
(2) Iron ferrite bars or rods (such as those used in radio antennas)
(1) Assortment of steel or iron rods or bars
(1) Assortment of nonmagnetic materials such as plastic and aluminum
(1) Soldering gun

Procedure

1. Identification of magnetic poles and forces
   a. Place the compass on your work surface. Arrange it so that it is as far as practical from the magnets and other magnetic materials. Note the direction indicated by the north-seeking end of the compass needle. Determine if the direction is approximately correct. If it is not, the compass may be in the magnetic field of some object in the room. Ask your instructor for assistance if necessary. What is the polarity of the north-seeking end of the compass needle?
   b. Bring the north pole of one of the bar magnets near the north-seeking end of the compass needle. Is the force that results one of attraction or repulsion?
   c. Bring the south pole of one of the bar magnets near the north-seeking end of the compass needle. What type of force is evidenced?
   d. Use both bar magnets. What force results when opposite poles are brought near one another? For like poles?
   e. Select a steel rod that is several inches long. Stroke the steel with a bar magnet as shown in Fig. 10-1. Repeat several times. Use the compass to identify the magnetic polarity of end A. Now, stroke the steel with the bar magnet, reversing the motion shown in Fig. 10-1. What is the polarity of the rod at end A?

2. Magnetic materials
   a. Place the compass on your work surface as far away from magnetic materials as is practical. Align one of the bar magnets so that it is perpendicular to the compass needle. Slowly approach the compass with the magnet until the needle deflects about 45° from its original position. Place the magnet on the work surface to maintain the position of the compass needle. Now, place lengths of steel and/or iron between the magnet and the compass. What happens to the needle deflection as the magnetic materials are placed on the work surface? Why?

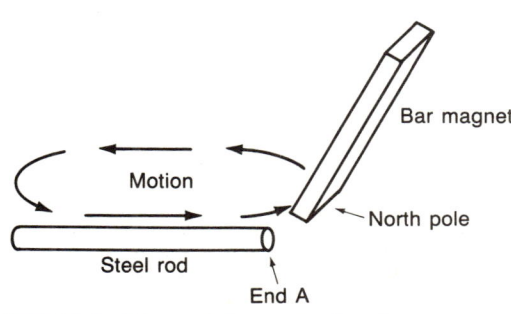

FIGURE 10-1 Magnetizing a steel rod.

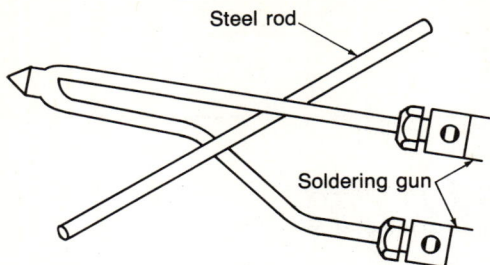

FIGURE 10-2 Demagnetizing a steel rod.

b. Remove the added magnetic materials but do not move the compass or the bar magnet. Now, place nonmagnetic materials (glass, plastic, wood, etc.) between the compass and the magnet. What happens to the deflection of the compass needle? Why?
c. Is there any attraction between the bar magnet and the nonmagnetic materials?
3. Temporary and permanent magnetism
a. Is there a force of attraction between the bar magnet and the ferrite rods?
b. Stroke the ferrite with a bar magnet as you did in step 1e. Test the ferrite with the compass. Does it develop and retain poles as the steel rod did? Is iron ferrite a temporary or a permanent magnet?
c. Magnetize a length of steel rod by stroking it with a bar magnet. Test it with the compass to verify that it retains poles. Now, slowly pass it through the tip of a soldering gun as shown in Fig. 10-2. The gun must be on, so be careful not to burn your fingers. Retest the steel rod with the compass. What has happened?

ACTIVITY 10-2   ELECTROMAGNETISM

Introduction

There is a magnetic field around any current-carrying conductor. This activity investigates a few of the principles of electromagnetism. The procedures are qualitative rather than quantitative for the reasons given in the last introduction.

Supplies

(2) Alkaline D cells
(1) Power supply, 0- to 25-V dc
(1) 6 ft of no. 22 solid wire (insulated)
(1) SPST momentary contact switch
(1) Roll of tape, transparent or masking
(1) Iron ferrite bar or rod
(1) Relay, 12-V dc with any contact arrangement (SPST, DPDT, etc.)
(1) Magnetic compass
(1) VOM or DMM

Procedure

1. Current-carrying conductors
a. Place the compass on your work surface as far as practical from any magnetic materials. Arrange a wire perpendicular to the compass needle and secure it with tape as shown in Fig. 10-3. Note that the wire is on top of the compass. Do not hold the switch closed any longer than necessary to properly observe the reaction since the load on the cell is very high. Close the switch. The needle should rotate 180°. If it does not, try two alkaline cells in series. The direction of flux, as viewed in Fig. 10-3, is from right to left under the wire. The needle will tend to move so that its own flux is in agreement with the flux around the wire. What rule predicts the flux around the wire? What happens when the cell is reversed? Why?
b. Change the arrangement to that shown in Fig. 10-4. Close the switch. What happens? Reverse the cell and repeat the procedure. What happens? Explain these results.
c. Change the arrangement to that shown in Fig. 10-5. Close the switch. What happens? Why?
2. Coils and cores
a. Wrap a few turns of wire around a ferrite core and place the coil near the compass as shown in Fig. 10-6. Close the switch and note the reaction of the needle. Increase the number of turns and do it again. Continue increasing the number of

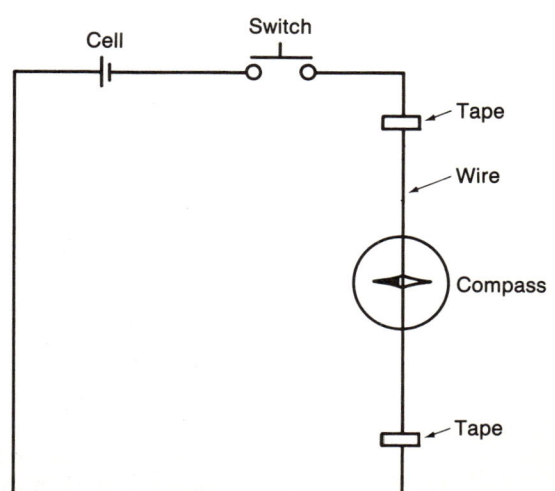

The north-seeking pole of the compass is black.

FIGURE 10-3 Detecting the magnetic field around a current-carrying conductor.

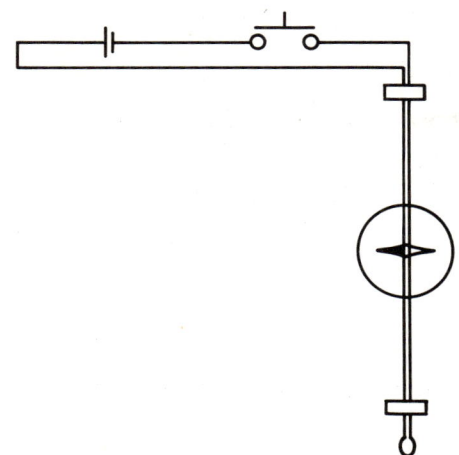

FIGURE 10-4 Electromagnetic field cancellation.

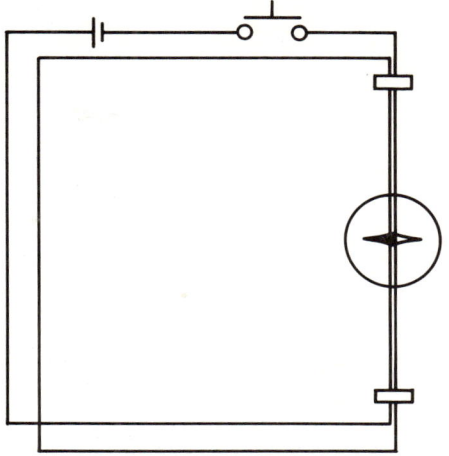

FIGURE 10-5 Electromagnetic field enhancement.

turns up to around 20. What can you conclude concerning the relationship between the flux strength of the coil and the number of turns? Where is the north pole of the coil as viewed in Fig. 10-6? Which rule predicts the magnetic polarity of the coil?

b. Remove the ferrite core from the coil. The coil should still have about 20 turns. Place it near the compass and close the switch. What has happened to the flux strength of the coil? Why?

3. Relay pick-up and drop-out current
   a. Construct a series circuit containing the dc power supply, the VOM or DMM set on a 100-mA current range, and the relay coil. Set the power supply for zero output. Slowly increase the output until the relay armature pulls in.

Note the current required for pick-up. Now, slowly decrease the power supply output until the armature returns to its open position. Note the current at the drop-out point. How do the two currents compare? Why?

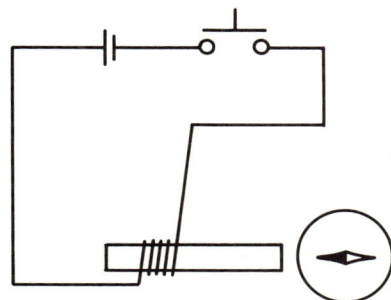

FIGURE 10-6 Using a core in an electromagnet.

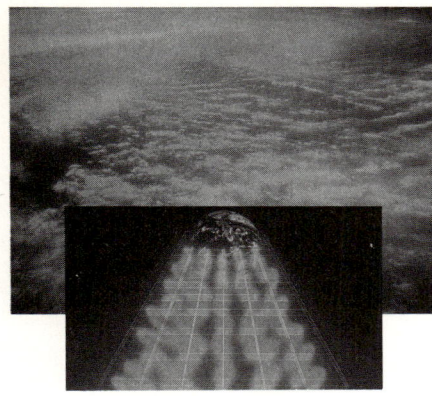

Chapter 11

Electrical Energy Sources

ACTIVITY 11-1 ELECTROMAGNETIC INDUCTION

Introduction

When a conductor moves across magnetic lines of force, an electromotive force is induced in the conductor. This activity investigates some of the principles associated with this method of generating electrical energy. Note: Meter movements can be damaged by a strong external magnetic field. Do not bring the bar magnets near the meter movement in this activity.

Supplies

(1) Panel meter, 100-μA dc (zero-center preferred but not necessary)
(2) Bar magnets
(1) Alkaline D cell
(1) Ferrite rod
(1) 10-ft length of no. 22 solid wire (insulated)
(1) 25-ft length of no. 28 magnet wire
(1) Knife
(1) Roll of tape
(1) SPST momentary contact switch
(1) Neon lamp, NE-2 or similar type
(1) Transformer, 12-V, dual 115-V primaries (Triad F107Z or equivalent)

Procedure

1. Electromagnetic induction
   a. Wind a coil using the no. 22 wire. Use two fingers as a form to shape the coil so that both bar magnets (placed face to face) can comfortably pass through the coil. Wind about eight turns and then start another layer but do not reverse the direction of the turns. When all of the wire is on the coil, remove it from your fingers and secure it with tape. Use a knife to scrape the insulation coating off the ends of the coil. Connect the coil to the meter movement and insert the bar magnet as shown in Fig. 11-1. Does the meter show current flow? Does the meter show current flow when the magnet is in the coil but not moving? Does the meter show current flow when the magnet is withdrawn from the coil? What can you conclude regarding motion and the induced current, disregarding polarity for the moment?
   b. Compare the meter deflection (polarity) for both directions of motion. What can you conclude?
   c. Reverse the bar magnet so that the south pole enters the coil first. Compare the meter deflection (polarity) for both directions of magnetic field. What can you conclude? What rule can be used to predict the induced polarity?
   d. Compare the meter deflection (magnitude) for both fast and slow motion. What can you conclude?
   e. Place the bar magnet in the coil. Rotate the bar about its long axis. Is there any significant meter deflection? Why?
   f. Place two bar magnets face to face with the north pole of one touching the south pole of the

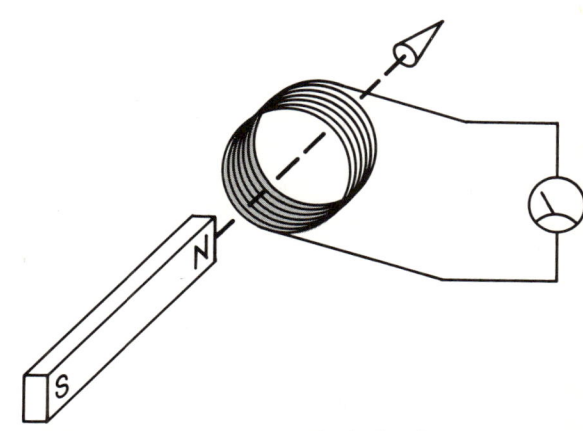

FIGURE 11-1 Electromagnetic induction.

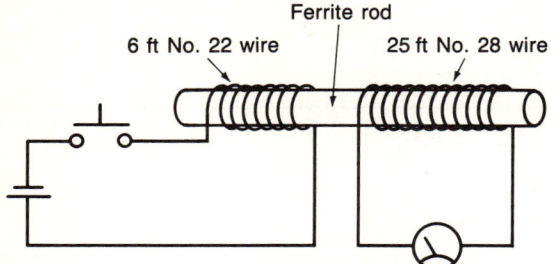

FIGURE 11-2 Experimenting with transformer action.

other. This combination decreases the external flux density. Thrust the magnets into the coil and note the magnitude of the meter deflection. Reverse one of the magnets so that its north pole touches the north pole of the other. This combination increases the external flux density. Thrust the magnets into the coil and note the magnitude of the meter deflection. What conclusion can you reach concerning flux density and the induced current?

g. Wind another coil using the no. 28 magnet wire. Use the same technique used to prepare the first coil. The length of the magnet wire is significantly greater, which will result in a coil with more turns. Connect the coil to the meter and compare its performance to the first coil. What kind of a relationship is there between the number of turns and the induced current?

2. Transformer action
   a. Build the circuit shown in Fig. 11-2. Close the switch while watching the meter. Open the switch while watching the meter. When does the meter deflect? Why?
   b. Is the polarity of the induced current the same when the switch is closed as when the switch is opened? Why?
   c. Reverse the alkaline cell and repeat the procedure. Is there any difference in the meter response? Why?
   d. A neon lamp will not light unless its gas is ionized. Ionization requires about 65 V. Connect the alkaline cell to the neon lamp. Does it light? Why?
   e. The voltage produced by electromagnetic induction is directly proportional to the number of turns. Figure 11-3 shows a transformer that has many turns of wire between terminals 1 and 4. Build this circuit. Press and release the switch. What is the response of the neon lamp?

ACTIVITY 11-2   CHEMICAL SOURCES OF ELECTRICITY

Introduction

When dissimilar metals are placed in an electrolyte solution, a difference in electrical potential is produced across the metals. This is the principle of operation for cells and batteries.

Supplies

(1) Index card, 3 × 5 in
(1) Piece of sheet copper, 1 × 3 in (circuit board stock is acceptable)
(1) Piece of sheet zinc, 1 × 3 in (galvanized steel is acceptable)
(1) VOM or DMM
(1) Steel-wool pad
(1) Dish of vinegar (standard 5% acetic acid content)

Procedure

1. Testing a cell
   a. Cut a piece of index card to a size slightly larger than the copper and zinc strips. Soak the index card in vinegar for about 5 min. Clean the surfaces of the copper and zinc strips with steel wool. Remove any remaining steel particles. Construct a cell as shown in Fig. 11-4. Pressure contacts are adequate. Soldering is not necessary. Measure the cell's output voltage.

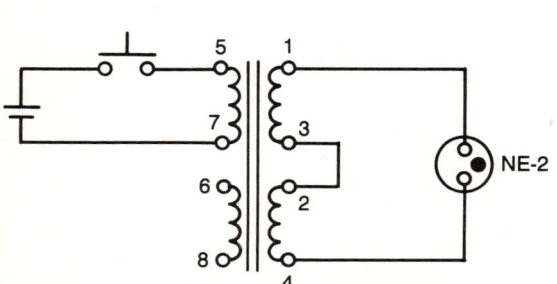

FIGURE 11-3 Voltage step-up in a transformer.

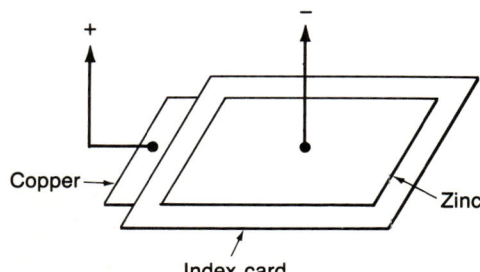

FIGURE 11-4 An experimental electrochemical cell.

b. Switch your meter to a 20-mA dc current range. Measure the cell's short-circuit current. How long can the cell provide maximum current?

c. Rest the cell for a few minutes. Remeasure the short-circuit current. Is it restored by resting?

d. Rest the cell for a few minutes. Slide the zinc strip to a position where it makes significantly less contact with the index card. Remeasure the cell voltage. Is there any significant change?

e. Remeasure the cell's short-circuit current. Is there any significant change? What can you conclude concerning the effect of the active area of the electrodes in an electrochemical cell?

NAME _____   DATE _____

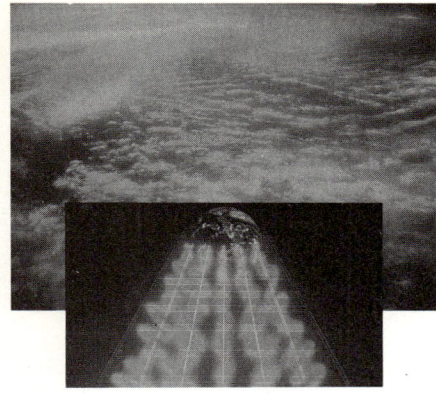

Chapter 12

Alternating Current

ACTIVITY 12-1 ALTERNATING CURRENT MEASUREMENTS

Introduction

Sinusoidal alternating current is constantly changing in amplitude and periodically reversing in polarity. Several different techniques of measurement are needed to describe ac waveforms completely. This activity deals with several of these techniques.

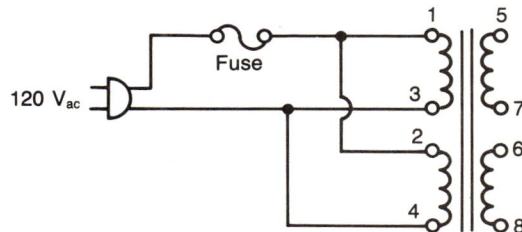

FIGURE 12-1 Transformer connection diagram.

Supplies

- (1) Transformer, dual 115-V primaries, dual 12-V, 2-A secondaries (Triad F107Z or equivalent)
- (1) Line cord assembly
- (1) Fuse, 1-A, 3-AGC with holder
- (2) Lamps, 14-V, 0.08-A, no. 756 (or equivalent) with holders
- (1) VOM or DMM
- (1) Oscilloscope
- (1) Power supply, 0- to 25-V dc

Procedure

CAUTION: The 120-V ac line used in this activity can cause a fatal shock. Use safe construction practices in connecting this circuit. Do not touch any part of the circuit after power is applied.

1. Voltage measurements
   a. Construct the circuit shown in Fig. 12-1 and apply power. Set up your multimeter to measure ac rms volts (20-V range). Measure the voltage from terminal 5 to terminal 7 of the transformer. Record this reading in the first row of Table 12-1. Measure the voltage from terminal 6 to terminal 8 of the transformer. Is it the same as the first reading?

   b. Refer to your oscilloscope manual and instructor for specific operating instructions. The following is offered as a guide to help you get started using the oscilloscope:

Power	On
Position controls	Centered
Brightness and focus	Centered
Volts/div	1 (with ÷10 probe)
AC/gnd/dc	AC
Sweep	Main (channel A)
Time/div	2 ms
Magnifier	Off
Triggering	Auto
Trigger selector	Internal
Trigger level	Centered

TABLE 12-1

	Calculated	Measured
V_{rms}	✕	
$V_{p\text{-}p}$		
V_P		
V_{av}		✕

Copyright © 1993 by the Glencoe Division of Macmillan/McGraw-Hill School Publishing Company. All rights reserved.

Connect your oscilloscope ground and the probe tip to terminals 5 and 7 of the transformer. If your oscilloscope is equipped with a ÷10 probe, each vertical division on the screen will be 10× the setting of the VOLTS/DIV selector. Complete Table 12-1. You should obtain good agreement between the calculated and measured values.

c. Adjust the dc power supply for the same voltage (V_{rms}) recorded in the first row of Table 12-1. Connect the power supply to one of the no. 756 lamps. Now, connect the second lamp to terminals 5 and 7 of the transformer. How does the brightness of the two lamps compare? Why?

ACTIVITY 12-2 PHASE

Introduction

Alternating current sources can be in phase or out of phase. This activity demonstrates some of the effects that phase angle can have on ac circuits.

Supplies

(1) Transformer, dual 115-V primaries, dual 12-V, 2-A secondaries (Triad F107Z or equivalent)
(1) Line cord assembly
(1) Fuse, 1-A, 3-AGC with holder
(2) Lamps, 14-V, 0.08-A, no. 756 (or equivalent) with holders
(1) Switch, SPST
(1) VOM or DMM
(1) Resistor, 560-Ω, $\frac{1}{2}$-W, ±5%
(1) Resistor, 1-kΩ, $\frac{1}{2}$-W, ±5%
(1) Resistor, 2.2-kΩ, $\frac{1}{2}$-W, ±5%

Procedure

CAUTION: The 120-V ac line used in this activity can cause a fatal shock. Use safe construction practices in connecting this circuit. Do not touch any part of the circuit after power is applied.

1. Phase
   a. Construct the circuit shown in Fig. 12-2(a). Close the switch and apply power. What is the condition of the two lamps? Measure the rms voltage from point A to point B. Open the switch. What is the condition of the two lamps? Measure the rms voltage from point A to point B.
   b. Power down. Change the circuit to that shown in Fig. 12-3(a). Close the switch and apply power. What is the condition of the two lamps? Measure the rms voltage from point A to point B. Open

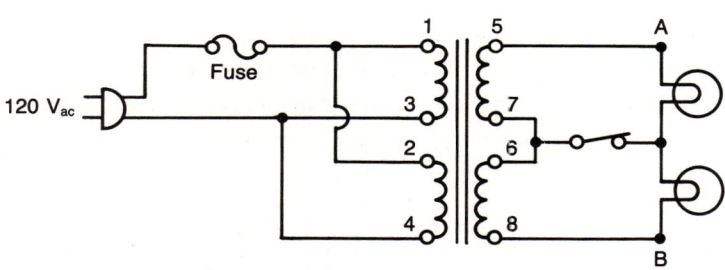
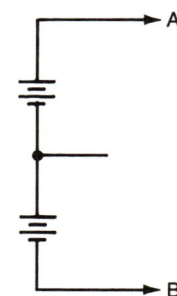

(a) Experimental circuit   (b) Analogous dc circuit

FIGURE 12-2 Transformer windings in-phase.

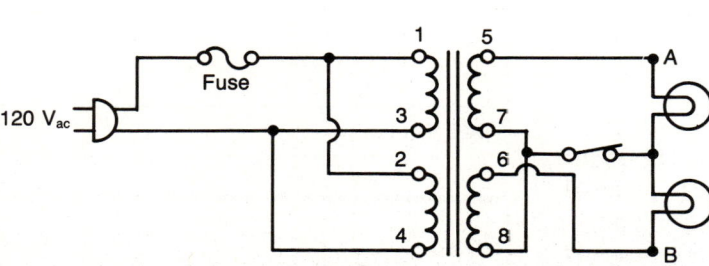
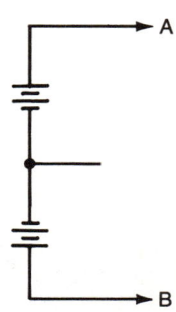

(a) Experimental circuit   (b) Analogous dc circuit

FIGURE 12-3 Transformer windings out-of-phase.

TABLE 12-2

	Calculated	Measured
V_{R_1}		
V_{R_2}		
V_{R_3}		
I_{R_1}		
I_{R_2}		
I_{R_3}		
P_{R_1}		
P_{R_2}		
P_{R_3}		

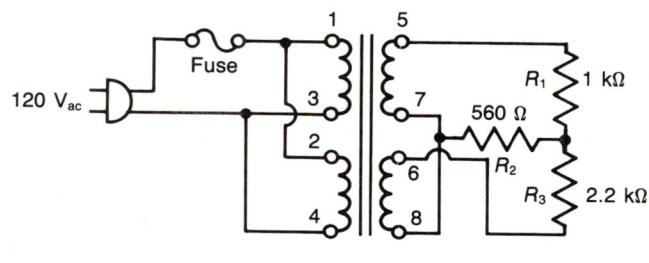

(a) Experimental circuit

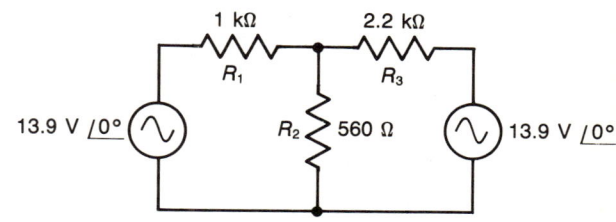

(b) Equivalent circuit

FIGURE 12-5 Modified phasor circuit.

the switch. What is the condition of the two lamps? What is the difference between Fig. 12-2(a) and Fig. 12-3(a)?

c. Power down. Build the circuit shown in Fig. 12-4(a). Apply power. Measure the voltage drops across the three resistors and record your results in Table 12-2. If your DMM has the capability to measure ac current, then also measure the three resistor currents and record the values in Table 12-2. Calculate the values for voltage, current, and power and record your results in Table 12-2. Figure 12-4(b) will assist you. The super-

position theorem is suggested. There should be good agreement between measured and calculated values.

d. Power down. Change the circuit to that shown in Fig. 12-5(a). Apply power and complete Table 12-3. There should be good agreement between measured and calculated values.

e. Compare the data in Tables 12-2 and 12-3. For example, why is the current in R_2 greater in Table 12-3?

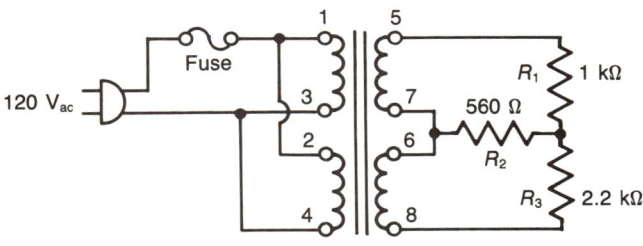

(a) Experimental circuit

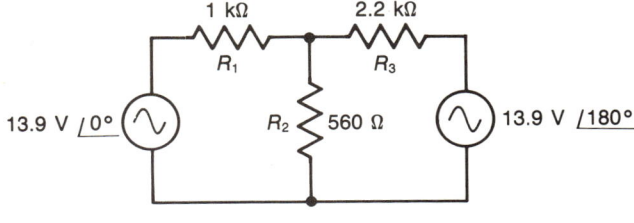

(b) Equivalent circuit

FIGURE 12-4 Phasor circuit.

TABLE 12-3

	Calculated	Measured
V_{R_1}		
V_{R_2}		
V_{R_3}		
I_{R_1}		
I_{R_2}		
I_{R_3}		
P_{R_1}		
P_{R_2}		
P_{R_3}		

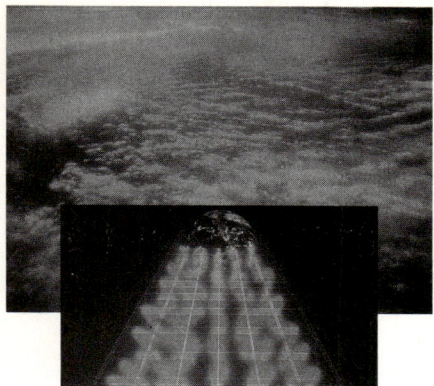

Chapter 13

Inductance

ACTIVITY 13-1 INDUCTORS AND MUTUAL INDUCTANCE

Introduction

Instruments that measure inductance are available but are not commonplace. This activity has been designed to observe the effect of inductance without actually measuring it. Inductance is that circuit property that opposes any change in current. In a sinusoidal ac circuit, the current is constantly changing. An inductor will therefore oppose the flow of sinusoidal current. This opposition is called *inductive reactance* and will be covered in detail in the next chapter. For this activity, it is important to know that the opposition is proportional to inductance. For example, a 2-H inductor offers twice the opposition as a 1-H inductor.

Supplies

- (2) Transformers, dual 115-V primaries, dual 12-V, 2-A secondaries (Triad F107Z or equivalent)
- (2) Lamps, 14-V, 0.08-A, no. 756 (or equivalent) with holders
- (1) VOM or DMM
- (1) Power supply, 60-Hz sinusoidal ac, 14-V
- (1) Roll of insulating tape

Procedure

◆ **CAUTION:** The transformers will be used as inductors in this experiment; however, they can produce a lethal voltage at terminals 1, 2, 3, and 4. These terminals will not be used in this experiment. Put insulating tape on them and do not touch them after power is applied.

1. Inductors in ac circuits
   a. If your ac supply is adjustable, use the meter and set it for 14 V. Connect a lamp to the supply and note its brightness. Turn off the supply and construct the circuit shown in Fig. 13-1. Turn on the power and record the brightness of the lamps. Also, measure the voltage across the lamps.
   b. Power down and add a second inductor (transformer) in series as shown in Fig. 13-2. Turn on the power and record the brightness and voltage drop across the lamps. Calculate the total inductance of this circuit assuming that each inductor is 9.32 mH and the mutual inductance is 0. Does an increase in inductance cause the opposition to alternating current to increase? Turn off the power.
   c. Change the circuit as shown in Fig. 13-3. Turn on the power and record the brightness and voltage drop across the lamps. Calculate the total inductance. Does a decrease in inductance cause the opposition to alternating current to decrease? Turn off the power.
   d. Remove the second transformer and build the circuit shown in Fig. 13-4. Record the brightness and voltage drop across the lamps. Calculate the total inductance assuming a mutual inductance of 8.78 mH (the fields are aiding in this circuit). Also, calculate the coefficient of coupling. Does the coefficient of coupling between the two windings approach unity?
   e. Power down and change the circuit as shown in Fig. 13-5. Apply power and record the brightness and voltage drop across the lamps. Calculate the total inductance (the fields are now opposing).

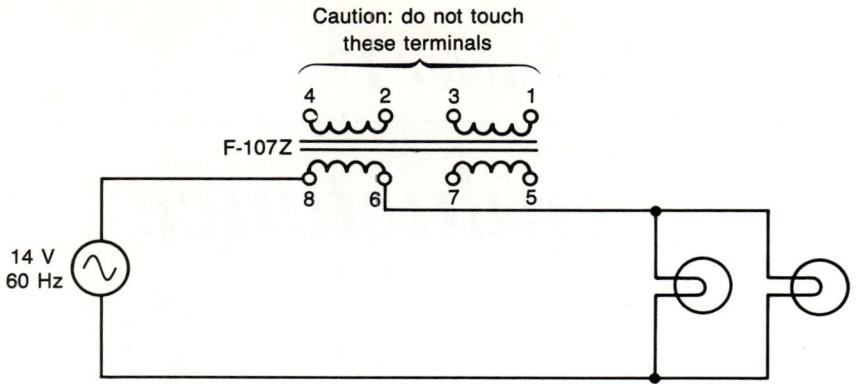

FIGURE 13-1 Inductor circuit.

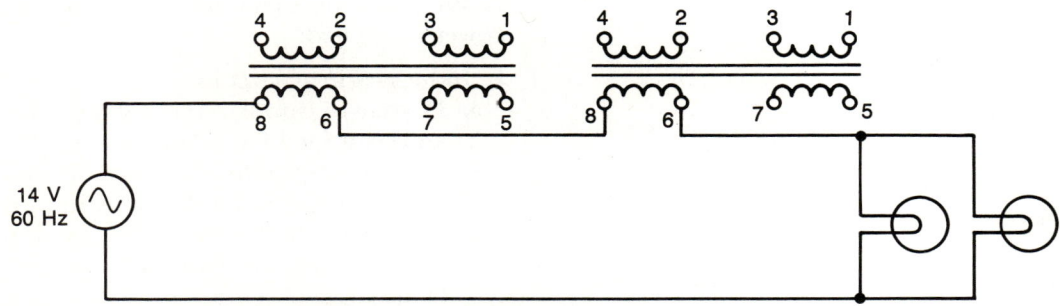

FIGURE 13-2 Inductors in series with no mutual inductance.

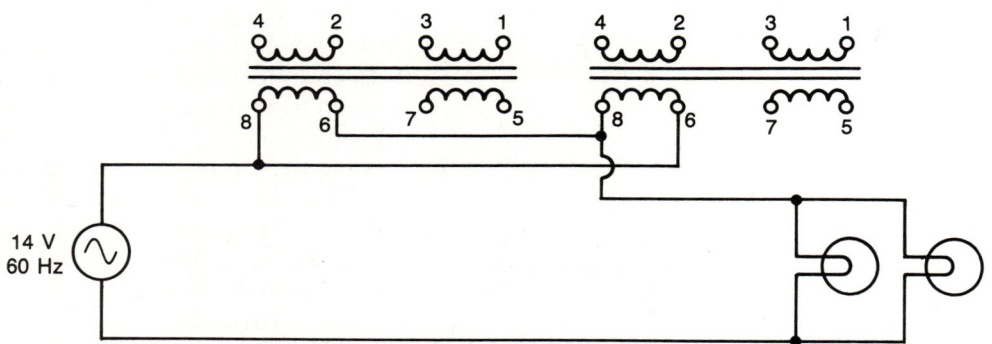

FIGURE 13-3 Inductors in parallel with no mutual inductance.

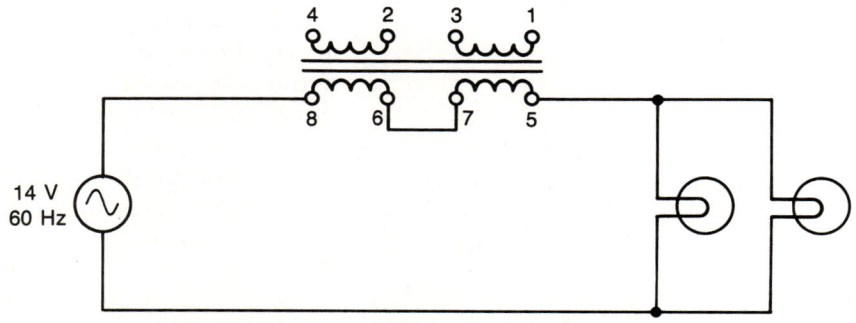

FIGURE 13-4 Inductors in series with mutual inductance (aiding).

50   Copyright © 1993 by the Glencoe Division of Macmillan/McGraw-Hill School Publishing Company. All rights reserved.

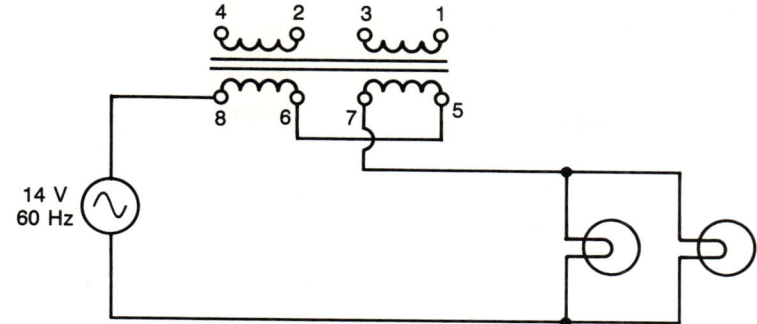

FIGURE 13-5 Inductors in series with mutual inductance (opposing).

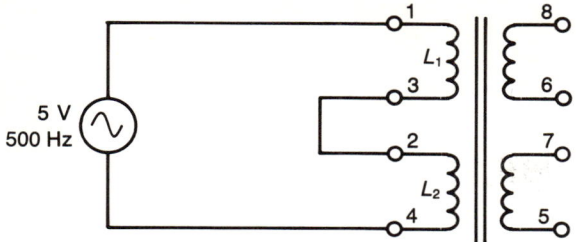

FIGURE 14-10 Circuit for determining L_A.

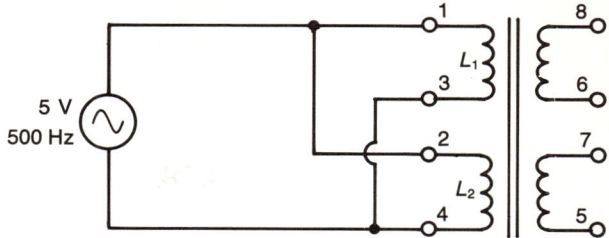

FIGURE 14-12 Parallel inductors with mutual inductance.

   b. Construct the circuit shown in Fig. 14-10, measure and record the current, and calculate and record X_{L_T} and L_A. (*Note:* L_A is the equivalent inductance of the series-aiding inductors.)

   c. Was your answer to step 2a correct?

3. Inductance of series-opposing inductors
   a. The 0.4-H inductor in Fig. 14-11 is included in the circuit so that the circuit current will not exceed the capacity of the generator.
   b. Construct the circuit shown in Fig. 14-11, and then measure and record the voltage between points A and B. This is the voltage used in calculating X_{L_T} for the series-opposing combination of L_1 and L_2. Next, measure the circuit current so that you can calculate and record values for X_{L_T} and L_0. (L_0 is the equivalent inductance of the series-opposing inductors L_1 and L_2.)
   c. Assuming that the generator could provide the required current while maintaining 5-V output, how much current would flow if the 0.4-H inductor had not been included in the circuit?

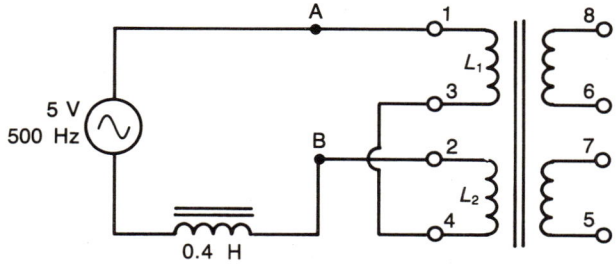

FIGURE 14-11 Circuit for determining L_0.

4. Mutual inductance and coefficient of coupling
   a. Using data accumulated so far, calculate and record the mutual inductance between L_1 and L_2.
   b. Now that you have determined L_1, L_2, and L_m, you can also calculate k (the coefficient of coupling). Record the value of k.
   c. What percentage of the flux produced by L_1 links L_2?

5. Parallel-aiding inductors
   a. If $k = 0$ for the circuit in Fig. 14-12, what would be the value of I_T? Use the data collected in steps 1a and 1b. Record this value of I_T. Because of the mutual inductance between L_1 and L_2, should the measured I_T be greater than or less than the value just recorded? Why?
   b. Construct the circuit shown in Fig. 14-12, and then measure and record I_T. Was your answer to the second question in step 5a correct?
   c. Why are the currents in steps 1a and 5b about equal?

ACTIVITY 14-4   RELATIONSHIP OF L, X_L, and f

Introduction

The first two activities in this chapter specified the circuits, components, and source conditions to be used, as well as a step-by-step procedure to be followed. This activity uses a different approach. It specifies the desired circuit characteristics, and you must design the circuit, select the components, and develop the procedure needed to verify your design experimentally.

Supplies

(1) Signal generator, audio range
(1) DMM with ac ranges
Miscellaneous components selected from the Materials List in this manual

Procedure

1. Design, construct, and test
   a. A circuit in which $I_T = 15.0$ mA and $X_L = 350\ \Omega$.
   b. A two-inductor circuit in which $I_{L_1} = 2.5 I_{L_2}$ and $I_T = 35.0$ mA.
   c. A series circuit in which $V_{L_2} = 1.7$ V, $V_T = 3.5 V_{L_1}$, and $I_T = 19$ mA.

2. Prepare a report which meets or exceeds the specifications given in Appendix 1.

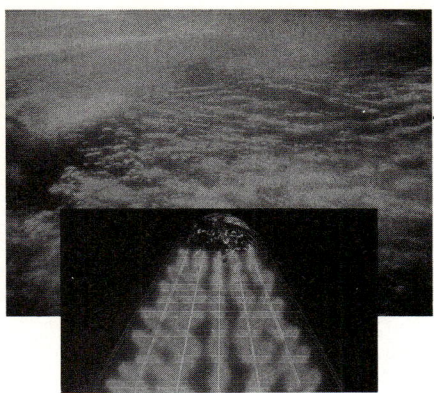

Chapter 15

Capacitance

ACTIVITY 15-1 DETERMINING CAPACITANCE

Introduction

The purpose of this activity is to determine experimentally the value of a capacitor by measuring voltage and indirectly measuring charge. Once charge and voltage have been determined, the capacitance can be determined because capacitance is defined as a constant of proportionality which relates voltage and charge.

Supplies

(1) Multimeter with high input resistance (≥ 10 MΩ) and a microampere current range
(1) Capacitor, 1000-μF, 50-V, $\pm 10\%$
(1) Resistor, 910-kΩ, $\frac{1}{2}$-W, $\pm 5\%$
(1) Power supply, 25-V dc
(1) Switch, SPST
(1) Watch with seconds indicator

Procedure

1. a. The capacitor in Fig. 15-1 is completely discharged. When S_1 is closed a current will start to flow. The instantaneous current will be V_R/R = 25 V/910 kΩ = 27.5 μA because $V_C = 0$ when the switch is first closed. This current will decay to zero by the time the capacitor is charged to 25 V. However, a few rough calculations show that it will take a long time for this current to decay. The charge required to charge the 1000-μF capacitor to 25 V is $Q = CV$ = 1000 μF $\times$ 25 V = 25,000 μC. Now, *if* the current continued at its 27.5 μA rate, the charge accumulating on the capacitor each second would be: $I = Q/t$; therefore, $Q = It$ = 27.5 μA $\times$ 1 s = 27.5 μC. After 1 min, the charge on the capacitor would only be 27.5 μC/s $\times$ 60 s = 1650 μC or less than 7% of the amount needed to charge the capacitor to 25 V. Of course, the current will be less than 27.5 μA after 60 s because some voltage (V_C) will be developed across the capacitor. The magnitude of this voltage would be $V = Q/C$ = 1650 μC/1000 μF = 1.65 V *if* the current remained constant at 27.5 μA. However, the current cannot remain constant because *if* $V_C = 1.65$ V, then $V_R = V_T - V_C$ = 25 V $-$ 1.65 V = 23.35 V and $I = V_R/R$ = 23.35 V/910 kΩ = 25.7 μA. Although all of these calculations are based on contradictory assumptions, they do show that after 1 min the voltage across the capacitor will be less than 1.65 V, and the circuit current will still be greater than 25.7 μA.

Athough the decay in current from 27.5 μA to zero follows an exponential curve, the section of the curve from 27.5 μA to approximately 26 μA can be approximated by a straight line. Thus, we will use the average of the circuit current in figuring the charge on the capacitor after 1 min of current flow.

↪ **CAUTION:** An electrolytic capacitor can explode if reverse polarity is applied to it! An exploding capacitor can be very hazardous. Not only are flying parts hazardous, but some electrolytes are quite caustic. Skin and clothing contaminated with electrolyte must be thoroughly washed with soap and water. In case of eye contact, flush open eyes with plenty of clean water for 15 min—seek immediate medical attention.

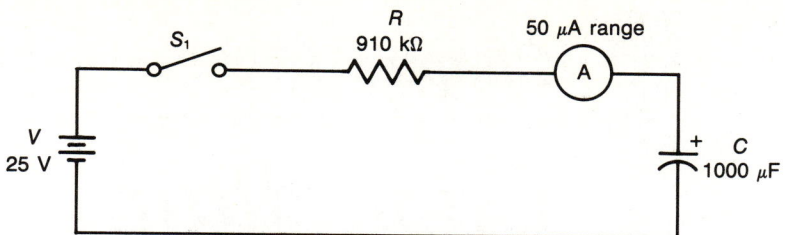

FIGURE 15-1 Circuit for determining capacitance.

	Trial		
	1	2	3
I_{Start}			
I_{End}			
V_C			
I_{av}			
Q			
C			

FIGURE 15-2 Data table for circuit of Fig. 15-1.

b. Construct the circuit in Fig. 15-1 with the switch open and the power supply turned off. Recheck the polarity of the electrolytic capacitor to be certain its negative lead is connected to the negative terminal of the power supply.

c. Check to be sure the capacitor is discharged by temporarily (1 min) shorting a jumper lead or test lead across its terminals. Remove this short and turn on the power supply. Now, while checking both your watch and the ammeter, close S_1 for exactly 1 min and note the current at the beginning and the end of the 1-min interval. *Be sure to open S_1 at the end of the 1-min interval.* Record the observed currents in the table in Fig. 15-2. Now measure, and record in Fig. 15-2, the voltage across the capacitor. Do not leave the voltmeter connected to the capacitor any longer than is necessary to obtain a reading because the voltmeter will slowly discharge the capacitor.

d. Using the currents and voltages you have just measured, calculate and record the remaining quantities called for under trial 1 in Fig. 15-2. The appropriate formulas are:

$$I_{av} = \frac{I_{start} + I_{end}}{2}$$
$$Q = I_{av}t$$
$$C = Q/V$$

Remember to keep track of units when using these formulas.

e. Within the tolerance limits of the components and equipment, does the value of C you determined in step 1d appear to be correct?

f. Sometimes the dielectric in an electrolytic capacitor is not too stable—especially if it has not been used for a while. Check your capacitor by repeating steps 1c and d two more times. Record the results in Fig. 15-2. Do your results on the three trials agree within 10 percent? If not, have your instructor help you reform the capacitor by charging it to 25 V and holding it at this voltage for a period of time.

g. Turn off the power supply. Now, check your understanding of what you have done by answering the following questions:

- If C in Fig. 15-1 were 2000 μF, what would the starting current be?
- What would be the approximate charge after 1 min if C were 4000 μF in Fig. 15-1?
- Approximately, how much voltage would you expect across C after 1 min if C were 2000 μF?

ACTIVITY 15-2 CAPACITOR INSULATION RESISTANCE

Introduction

When the insulation resistance of a capacitor is not high enough, a noticeable current (leakage current) flows through the dielectric. When a capacitor with high leakage current is in series with a large resistance, an undesired voltage drop can appear across the resistor. This activity will compare electrolytic and film capacitors in terms of their leakage current (insulation resistance).

Supplies

(1) DMM with 0.1-μA resolution
(1) Capacitor, 1000-μF, 50-V, ±10%
(1) Capacitor, 500-μF, 50-V, ±10%
(1) Capacitor, 0.1-μF, 25-V, ±10%
(1) Capacitor, 0.01-μF, 25-V, ±10%
(1) Power supply, 0- to 25-V dc

Procedure

CAUTION: Electrolytic capacitors can explode when the polarity of the applied voltage is reversed. Always double check the polarity before applying power. If in doubt, check with your instructor.

1. **Leakage Current**
   a. Set the power supply for zero output. Construct the circuit shown in Fig. 15-3. Use the highest current range of your meter and the 1000-μF capacitor. Recheck the capacitor polarity and then turn on the power supply and increase its output to 25 V. Reduce the ammeter range until a reading is obtained. Leave the circuit connected and monitor the ammeter reading until the reading has either been stable for 1 min or until the last digit on the DMM vacillates up and down. This is the value of the leakage current for the 1000-μF capacitor you are using. Record this value in Fig. 15-4. Using this current and the source voltage, compute the insulation resistance for the capacitor. Also, record this value in Fig. 15-4.
   b. Without turning off the power supply or reducing its output voltage, disconnect the ammeter from the circuit. Now, reduce the power supply output to zero. How much voltage should be on the capacitor at this time? Use the DMM to measure the voltage across the 1000-μF capacitor. Is your answer to the above question correct? What should happen if you use a test lead or a jumper wire and short the capacitor leads together? Try it. Leave the capacitor discharged.
   c. Using the procedure specified in step 1a, measure, calculate, and record the values needed to complete Fig. 15-4.
   d. Was there a measureable leakage current through the foil capacitors?
   e. From the data you have collected, can you conclude that the dielectric resistance of a foil capacitor is infinite? Why?

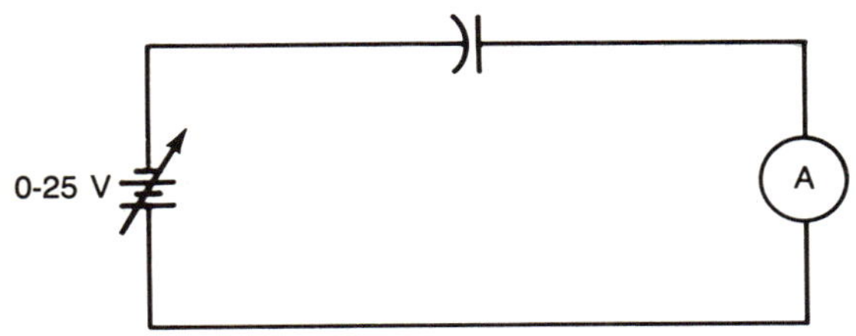

FIGURE 15-3 Circuit for testing leakage current.

Capacitance	Leakage current	Insulation resistance
1000μF		
500μF		
0.1μF		
0.01μF		

FIGURE 15-4 Data table for circuit of Fig. 15-3.

Chapter 16

Capacitors in DC and AC Circuits

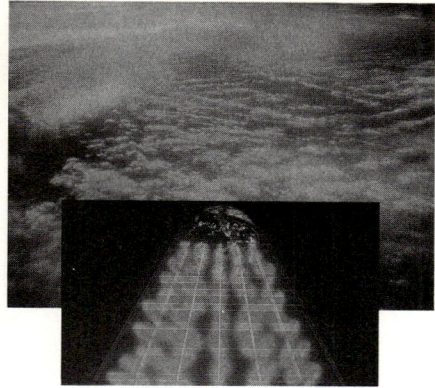

ACTIVITY 16-1 RC TIME CONSTANTS

Introduction

The effects of charging and discharging time on capacitor voltage will be studied in this activity. Both the oscilloscope and the DMM will be used to measure and observe the voltage changes in an RC circuit.

Supplies

(1) Oscilloscope with an ×10 probe and a 2 s/div. sweep
(1) DMM
(1) Signal generator, audio range with square-wave function
(1) Power supply, 0- to 25-V dc
(1) Capacitor, 1-μF, 50-V, ±10%
(1) Capacitor, 0.1-μF, 50-V, ±10%
(1) Resistor, 10-kΩ, $\frac{1}{2}$-W, ±5%
(1) Switch, SPST
(1) Watch with a seconds indicator

Procedure

1. Observing a discharge curve on the oscilloscope
   a. Connect the ×10 probe to the oscilloscope. With this probe, the input resistance of most oscilloscopes will be 10 MΩ. Measure and record the input R of the oscilloscope you are using by measuring the resistance between the probe input leads.
   b. The above-recorded resistance will be the resistance through which you will charge and discharge a 1-μF capacitor. What will be the time constant (τ) of this RC circuit? The capacitor, as shown in the Fig. 16-1, will be charged to 20 V before being discharged through the internal resistance of the oscilloscope. Using the formula $V_{Ct} = V_{CE}e^{-t/\tau}$, calculate and record the voltages required to complete the first column in Fig. 16-2.
   c. Adjust the oscilloscope controls to meet the following requirements:

   - DC-coupled vertical input calibrated for 0.5 V/div. (5 V/div. including the ×10 probe).
   - Horizontal sweep free-running and calibrated for 2 s/div. This provides a dot which moves slowly across the screen (one division each 2 s).

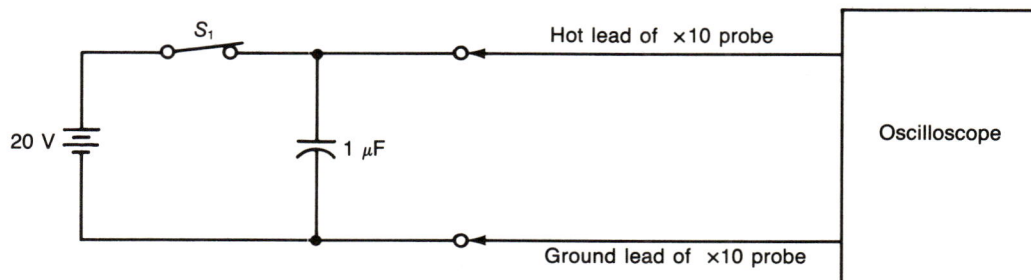

FIGURE 16-1 Observing the discharging of a capacitor.

Copyright © 1993 by the Glencoe Division of Macmillan/McGraw-Hill School Publishing Company. All rights reserved.

Capacitor discharging		
Number of time constants	Voltage across capacitor	Discharging time (measured)
0	20 V	0 s
0.5		
1.0		
1.5		
2.0		
2.5		

FIGURE 16-2 Data table for discharging capacitor.

Capacitor charging			
Number of time constants	Voltage across oscilloscope V_R	Voltage across capacitor V_C	Charging time (measured)
0	20 V	0 V	0 s
0.5			
1.0			
1.5			
2.0			
2.5			

FIGURE 16-4 Data table for charging capacitor.

- Sweep line located on the bottom graticule line and starting to the left of the left-most graticule line.

With the oscilloscope calibrated as above, what voltage and time are represented when the moving dot is two divisions up and two divisions right of the lower left corner of the graticule?

d. Construct the circuit shown in Fig. 16-1. This circuit allows you to charge a 1-μF capacitor to 20 V and then, by opening S_1, discharge it through the internal resistance of the oscilloscope. The oscilloscope allows you to measure the time required (after S_1 was thrown to position 2) for the capacitor to discharge to a specified voltage. If the oscilloscope is properly calibrated, the dot should be sweeping four divisions above the bottom graticule line.

Open S_1 at the instant the moving dot on the oscilloscope is at the first vertical graticule mark and observe how many horizontal divisions the dot moves as its vertical position moves to a height which represents the capacitor voltage you calculated for 0.5 τ. At 2 s/div., how much time did it take to discharge to this voltage? Based on the value of τ calculated in step 1b, is this the time you expected?

e. By opening and closing S_1 (at the appropriate time!), measure the times required to discharge to the other voltages recorded in Fig. 16-2. Record these times in the second column of Fig. 16-2. Are these the times you expected?

2. Observing a charge curve on the oscilloscope
   a. The circuit in Fig. 16-3 will allow you to observe the effects of a charging capacitor in an RC circuit. In this circuit, the oscilloscope responds to the instantaneous voltage across its internal resistance (which is the R for the RC time constant). Kirchhoff's voltage law tells us that $V_T = V_{R,\text{inst}} + V_{C,\text{inst}}$, so determining the time for the resistor voltage to decay to 36.8 percent of V_T is the same as determining the time for the capacitor voltage to increase to 63.2 percent of V_T. For the circuit in Fig. 16-3, determine V_R and V_C for the time constants specified in Fig. 16-4. Record these voltages in Fig. 16-4.

   Compare the V_R column in Fig. 16-4 with the V_C column in Fig. 16-2. Are they the same? Why?

   b. You can measure the time required to charge the capacitor in Fig. 16-3 to any of the values listed in Fig. 16-4. Just open S_1 in Fig. 16-3 at the instant the moving dot crosses the first graticule

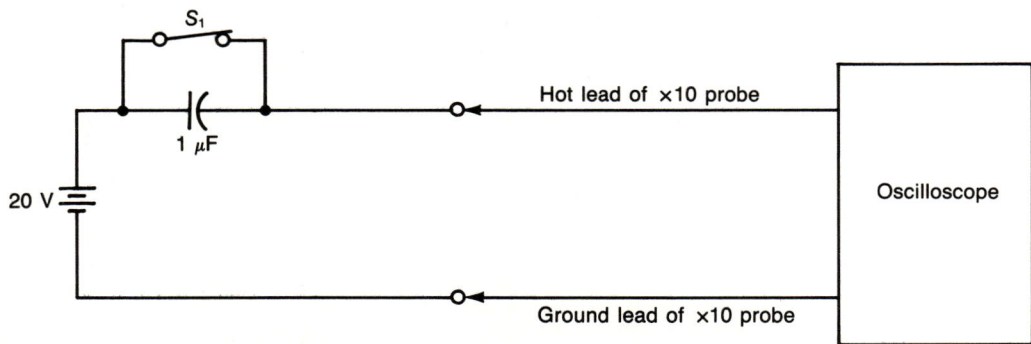

FIGURE 16-3 Observing the charging of a capacitor.

mark and record the time when the dot has decreased to a vertical level representing the desired voltage. Using this procedure, measure (and record in the table) the times needed to complete the table in Fig. 16-4. Do the times agree, within tolerances, with the values you would calculate from the value of τ?

c. The oscilloscope in Figs. 16-1 and 16-3 measured both time and voltage as well as providing a visual display of the charge-discharge curves of an RC circuit. However, neither the time nor the voltage could be measured with much accuracy.

If the oscilloscope in these figures is replaced with a DMM, the capacitor will charge and discharge through the DMM's internal resistance. The internal resistance of a DMM is also typically 10 MΩ. With the oscilloscope replaced by the DMM, the time to charge or discharge to a specified voltage must be measured with a second-indicating watch.

Replace the oscilloscope in Fig. 16-3 with the DMM and determine the time required to charge the capacitor to 17.3 V. Record this time. Does this time agree with the time determined by the oscilloscope method?

3. Discharging through a voltage source
   a. How long will it take the capacitor in Fig. 16-5(b) to reach 15 V after S_1 is closed?
   b. When V_C is 15 V, how much voltage will the DMM indicate?
   c. Will the capacitor be charging or discharging?
   d. After 90 s, what value of voltage will the DMM be indicating?
   e. Construct the circuit in Fig. 16-5(a) and close S_1 for several minutes so that the capacitor is completely charged. Then open S_1 and reduce the source voltage to 10 V. Now, time how long it takes the capacitor to discharge to 15 V. Why did the DMM indicate reverse polarity? Does the measured time agree with the time predicted in step 3a above? Leave S_1 closed until the capacitor is discharged to 10 V. Now open S_1.
   f. Without shorting out or discharging the capacitor, reverse its leads so that you have the circuit shown in Fig. 16-5(c). Now when S_1 is closed, the source and capacitor voltages will be aiding each other. How much voltage will be across the DMM the instant S_1 is closed? Will the capacitor be charging or discharging? How long will it take to charge the capacitor to 5 V of the opposite polarity to that shown in Fig. 16-5(c)? At this time, how much voltage will the DMM indicate? Close S_1 and check your answers to these questions.

4. Square-wave response
   a. What is the time constant of the circuit in Fig. 16-6?
   b. What is the time that a square wave will be positive (or negative) if the frequency is 100 Hz? 5 kHz?
   c. Would you expect the voltage across the resistor in Fig. 16-6 to be closer to a square wave when the frequency is 100 Hz or 5kHz? Why?
   d. Construct the circuit in Fig. 16-6 and adjust the generator for a 6-V_{p-p}, 5-kHz output. Observe the waveform of the voltage across the resistor. Now decrease the generator's frequency to 100 Hz and observe the voltage across the resistor. Draw this waveform in Fig. 16-7. Was your answer to step 4c correct?

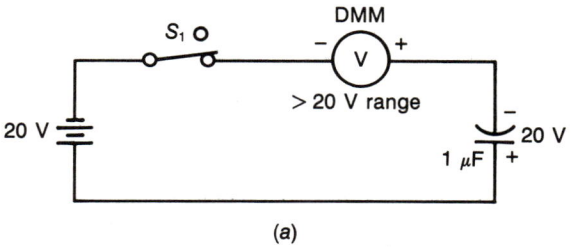

(a)

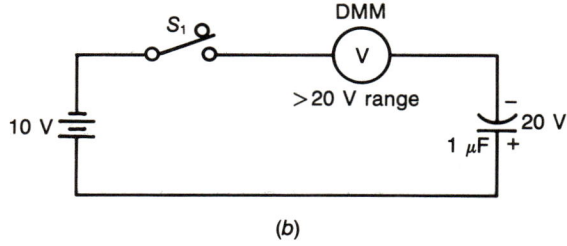

(b)

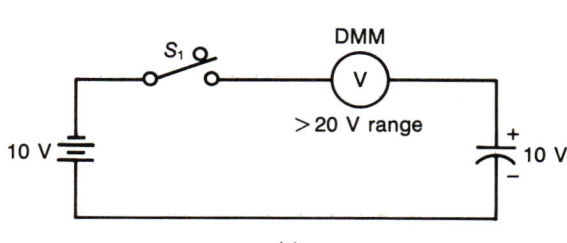

(c)

FIGURE 16-5 Circuits for step 3.

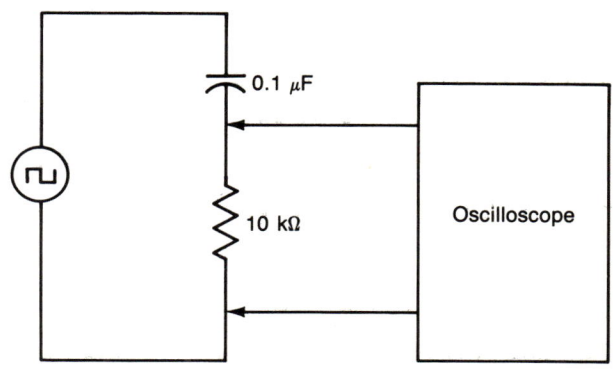

FIGURE 16-6 Circuit for observing the square-wave response of an RC circuit.

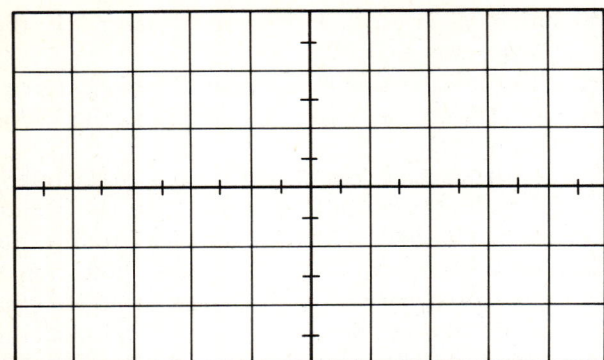

FIGURE 16-7 Graph for resistor voltage waveform at 100 Hz and 5 kHz.

ACTIVITY 16-2 CAPACITORS IN AC CIRCUITS

Introduction

After investigating the relationships between X_C, f, and C, this activity will demonstrate the behavior of capacitors in series, parallel, and series-parallel circuits. After predicting circuit values, the circuits will be constructed and measurements will be made to verify the predictions.

Supplies

(1) DMM with ac voltage and current ranges
(1) Signal generator, audio range
(1) Capacitor, 0.5-μF, 50-V
(2) Capacitors, 0.1-μF, 50-V
(1) Capacitor, 0.05-μF, 50-V

Procedure

1. X_C, f, and C relationships
   a. If C is doubled while V_C and f are held constant, what should happen to I_C in a capacitor circuit?
   b. Construct the circuit in Fig. 16-8 using the 0.5-μF capacitor and a frequency of 100 Hz. Measure I_C and V_C and record these values in the table in Fig. 16-9. Change the value of C to

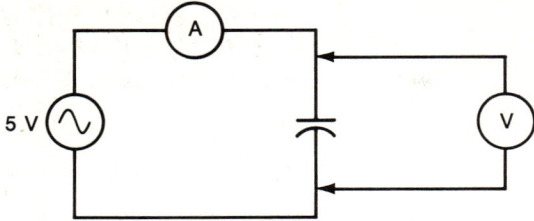

FIGURE 16-8 Circuit for measuring I_C and V_C for determining reactance.

0.1 μF and then to 0.05 μF to obtain the other values of I_C and V_C called for in the first row of Fig. 16-9. You may have to adjust the generator's output voltage after any change in the circuit, including a change in frequency. Always adjust the generator's output voltage after the circuit is completed. Do the measured currents support your answer to step 1a above? Use Ohm's law and your data to compute the reactances called for in row 1 of Fig. 16-9. What is the relationship between X_C and C?
   c. If the frequency is doubled while C and V_C are held constant, what should happen to I_C?
   d. Make the necessary circuit changes, measurements, and calculations needed to complete the table in Fig. 16-9. Do the results you obtained support your answer to step 1c?
   e. What is the relationship between X_C and f?
   f. From the data you have gathered, estimate and record the reactance of a 1-μF capacitor at 10 Hz. Now, using the formula $X_C = 1/(2\pi fC)$, calculate and record the reactance of a 1-μF capacitor at 10 Hz. Allowing for measurement error, does your estimate agree with your calculations?
   g. If you halved the source voltage and doubled the source frequency, what would happen to the source current in a capacitor circuit? Check your answer by doing so to the circuit in Fig. 16-8.

2. Series capacitor circuits
   a. For the circuit shown in Fig. 16-10, calculate and record I_{C_T}, X_{C_1}, X_{C_2}, X_{C_3}, X_{C_T} and V_{C_T}.
   b. Construct the circuit in Fig. 16-10 and make the measurements and calculations needed to complete Fig. 16-11. Record all measured data and calculated values.

	$C = 0.5\ \mu F$			$C = 0.1\ \mu F$			$C = 0.05\ \mu F$		
f	I_C	V_C	X_C (V_C/I_C)	I_C	V_C	X_C (V_C/I_C)	I_C	V_C	X_C (V_C/I_C)
100 Hz									
200 Hz									
400 Hz									

FIGURE 16-9 Data table for the circuit in Fig. 16-8.

c. Does changing the frequency of the source voltage change the voltage distribution in a series capacitor circuit?
d. In a series capacitor circuit, how is the source voltage distributed in relation to the size of the individual reactances and the individual capacitances?
e. Would shorting out C_1 or C_3 have the greater influence on the current in the circuit in Fig. 16-10? Why? Check your answer by momentarily shorting together the terminals of C_1 and then C_3.

3. Parallel capacitor circuits
   a. For the circuit in Fig. 16-12, calculate and record X_{C_T}, I_{C_T}, I_{C_1}, I_{C_2} and I_{C_3}.
   b. Construct the circuit in Fig. 16-12. Make and record the measurements and calculations needed to complete Fig. 16-13.
   c. What effect does halving the frequency in a parallel capacitor circuit have on X_T, I_T, and the current distribution?
   d. Would removing a capacitor from Fig. 16-12 cause I_T to increase or decrease? Removal of which capacitor would have the greatest effect? Why?
   e. How is I_T distributed in relation to the size of the individual capacitances in a parallel circuit?
   f. Do both the calculated and the measured currents for the circuit in Fig. 16-12 satisfy Kirchhoff's current law? Should they?

4. Series-parallel capacitor circuits
   a. For the circuit in Fig. 16-14, compute C_T, X_{C_T}, I_T, I_{C_3}, I_{C_4}, V_{C_1}, V_{C_2} and V_{C_4}.
   b. Construct the circuit in Fig. 16-14 and make the measurements and calculations needed to complete Fig. 16-15. If values in step 4a do not agree with these values within ±15 percent, recheck your work in both steps.
   c. Will shorting C_3 in Fig. 16-14 cause I_{C_4} to increase or decrease? Check your answer by shorting C_3 while measuring I_{C_4}.
   d. What will happen to V_{C_2} in Fig. 16-14 if the frequency is doubled? Is halved? Check your answer by changing the frequency, but be sure the source voltage remains at 5 V.

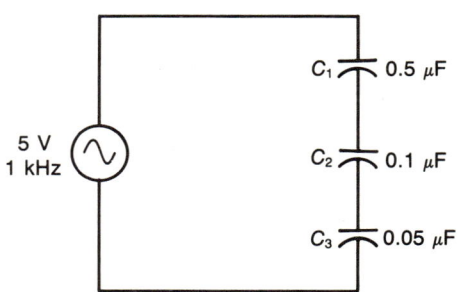

FIGURE 16-10 Series capacitor circuit.

V_T	I_T	V_{C_1}	V_{C_2}	V_{C_3}	X_{C_T} (V_T/I_T)	X_{C_1} (V_{C_1}/I_T)	X_{C_2} (V_{C_2}/I_T)	X_{C_3} (V_{C_3}/I_T)
5 V, 1 kHz								
5 V, 500 Hz								

FIGURE 16-11 Data table for series capacitor circuit.

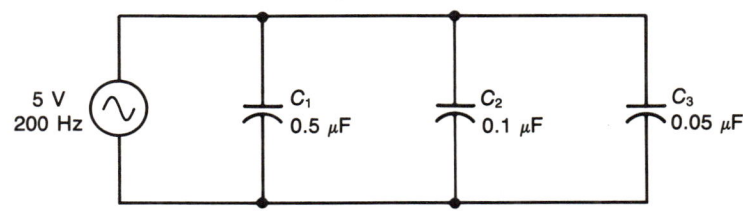

FIGURE 16-12 Parallel capacitor circuit.

V_T	I_T	I_{C_1}	I_{C_2}	I_{C_3}	X_{C_T} (V_T/I_T)	X_{C_1} (V_T/I_{C_1})	X_{C_2} (V_T/I_{C_2})	X_{C_3} (V_T/I_{C_3})
5 V, 200 Hz								
5 V, 100 Hz								

FIGURE 16-13 Data table for parallel capacitor circuit.

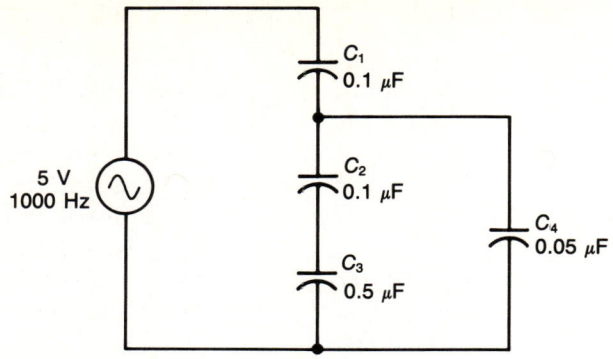

FIGURE 16-14 Series-parallel capacitor circuit.

V_T	I_T	I_{C_3}	I_{C_4}	V_{C_1}	V_{C_2}	V_{C_4}	X_{C_T} (V_T/I_T)
5 V							

FIGURE 16-15 Data table for series-parallel capacitor circuit.

ACTIVITY 16-3 SERIES-PARALLEL CAPACITOR CIRCUITS

Introduction

In this activity you are to select the components, circuit configurations, and procedures needed to produce the specified results. Rather than responding to specific questions about a given concept, you will write a detailed report to show that you understand the concepts involved in the activity.

Supplies

(1) Signal generator, audio range

(1) DMM with ac ranges

Miscellaneous capacitors selected from the Materials List in this manual

Procedure

1. Design, construct, and test a series-parallel capacitor circuit in which $I_{C_1} = 2.5 I_{C_2} = 2.5 I_{C_3} = 5 I_{C_4}$.
2. Design, construct, and test a series-parallel capacitor circuit in which $V_T = V_{C_1} + V_{C_2} + V_{C_3}$ and $I_T = I_{C_2} + I_{C_4} = 10$ mA and $I_{C_1} = 6 I_{C_2}$.
3. Write an activities report which meets or exceeds the requirements listed in Appendix 1.

Chapter 17

RCL Circuits

ACTIVITY 17-1   DIFFERENTIATION AND INTEGRATION

Introduction

In this activity, you will determine the characteristics of an *RL* differentiator circuit and an *RC* integrator circuit. A unijunction transistor (UJT) will be used in the integrator circuit to discharge the capacitor whenever the integrated voltage reaches a specified value.

Supplies

(1) Oscilloscope with ×10 probe
(1) Signal generator, audio range with square-wave function
(1) Power supply 0- to 25-V dc
(1) Diode, IN4002 or equivalent
(1) UJT, 2N2646 or equivalent
(1) Resistor, 1-kΩ, $\frac{1}{2}$-W, ±5%
(1) Capacitor, 0.1 μF, 25-V, ±10%
(1) Inductor (RF choke), 2.5-mH, <10-Ω dc resistance, ±10%

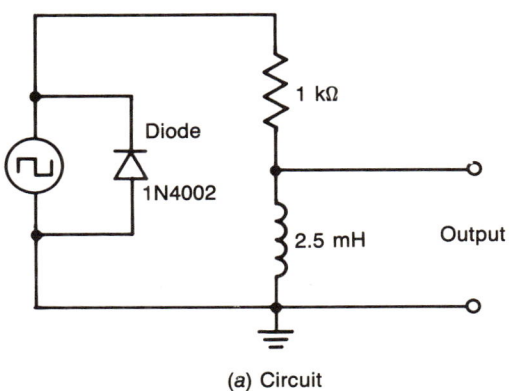

(a) Circuit

(b) 1N4002 lead connections

FIGURE 17-1 *RL* differentiator.

Procedure

1. Differentiator circuits
   a. An *RL* differentiator circuit is shown in Fig. 17-1. The diode in the circuit converts a square-wave ac output from the generator to a positive-going train of symmetrical pulses. If a generator provides positive-going pulses, this diode is not needed. Should this circuit provide more differentiation at 200 kHz or 5 kHz? Why?
   b. What is the τ for this circuit?
   c. Assuming that the CEMF of the inductor decays to zero in 5 τ, at what frequency should the CEMF in this circuit have just enough time to reduce to zero?
   d. For this circuit, should the peak-to-peak value of V_L be greater than, less than, or equal to the peak-to-peak value of V_T?
   e. Construct the circuit in Fig. 17-1 and adjust the generator's output, *when connected to the circuit*, for a 5-V$_{p-p}$, 200-kHz square wave. Use the calibrated oscilloscope to measure the generator's output voltage (V_T). Set the scope for dc coupling and the vertical gain for 2 V/div. Adjust the vertical position so that the zero vertical input is the vertical center of the graticule. Adjust the horizontal controls to display one cycle which spans eight graticule marks. Locate the fast-rising voltage spike on the second vertical graticule line. Now, measure the output (across the inductor) with the oscilloscope. Is the output fully differentiated? Is the output greater than 5 V$_{p-p}$? Draw the output waveform in Fig. 17-2. Don't change any of the vertical controls on the oscilloscope while working with this circuit.

Copyright © 1993 by the Glencoe Division of Macmillan/McGraw-Hill School Publishing Company. All rights reserved.

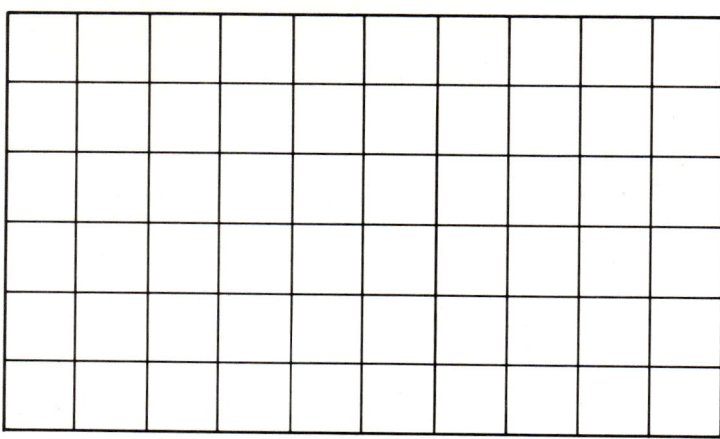

FIGURE 17-2 Graph for waveforms for the *RL* differentiator.

f. Change the generator frequency to 5 kHz and adjust its output for 5 $V_{p\text{-}p}$. Don't change any of the vertical controls on the oscilloscope, but re-adjust the horizontal for one cycle located as explained in step 1e. Now, measure the output voltage. Was your answer to step 1a correct? Is the waveform now sharply differentiated? Using a dashed line, also draw this observed waveform in Fig. 17-2.

g. Don't change the oscilloscope or the generator controls. Now interchange *R* and *L* so that *R* is connected to the grounded output lead of the generator. Next, observe the waveform across *R*. Describe this waveform. Is this the type of waveform one would predict by applying Kirchhoff's voltage law to the circuit?

h. Now, change the generator frequency to the frequency you determined in step 1c and adjust the output to 5 $V_{p\text{-}p}$. Adjust the scope for 1 cycle as explained in step 1e. Does the output waveform show the cemf reducing to zero at the end of each half cycle? Again, draw this waveform in Fig. 17-2, but use a dotted line.

2. Integrator circuit

a. The integrator circuit in Fig. 17-3 uses a UJT to discharge the capacitor when the capacitor's voltage reaches a specified level. We must know the rudimentary characteristics of the UJT to understand how the UJT periodically discharges the capacitor. In the explanation of UJT which follows, all voltages are referenced to the B_1 lead.

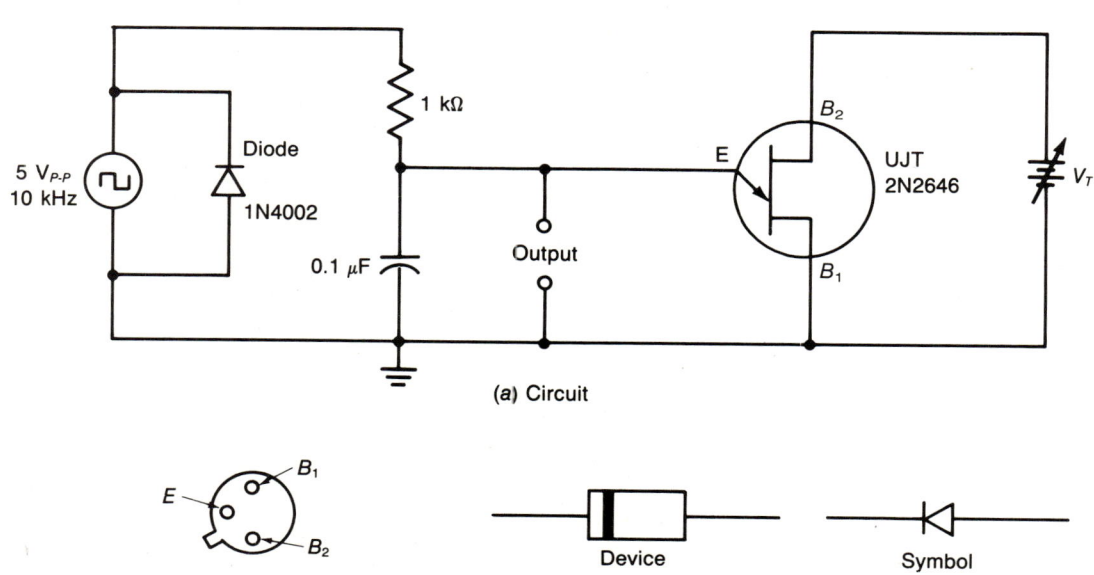

FIGURE 17-3 *RC* integrator.

The dc voltage applied to the UJT is uniformly distributed across the internal element between the B_1 and B_2 leads as long as the E lead is less than about 60 percent of the applied dc voltage. As long as the voltage on the E lead is less than 60 percent of V_T, the E lead is essentially insulated from both the B_1 and the B_2 leads. The resistance between E and B_1 or B_2 is hundreds of megohms. The only current flow in the UJT, at this time, is the current flowing from B_1 to B_2 through the resistance (about 7 kΩ) of the material internally connecting B_1 to B_2. When the voltage on E reaches about 60 percent of the voltage between B_1 and B_2 (V_T in this circuit), the resistance between E and B_1 suddenly drops to a few ohms and a large current flows between B_1 and E until the voltage on E drops to about 1.5 V. When current flows between E and B_1, the UJT is said to be "turned on." At about 1.5 V, the resistance between E and B_1 suddenly changes back to hundreds of megohms and current flow between E and B_1 stops. The UJT is turned off. No current flows between E and B_1 until the voltage on E again reaches about 60 percent of V_T.

Now we can see how the circuit in Fig. 17-3 operates. The capacitor integrates the +5-V pulses from the generator until its voltage is 60 percent of V_T. At that time, the UJT turns on and the capacitor discharges through the low E-B_1 resistance until the capacitor's voltage is less than the critical voltage (about 1.5 V) needed to keep the UJT turned on. Then the E-B_1 resistance jumps back to a very high value again, and the capacitor starts to recharge as it integrates the square-wave input to the RC integrator section of the circuit. This process of the capacitor charging to 60 percent of V_T and then discharging to 1.5 V is continuously repeated.

Questions: If V_T is +10 V and the input pulses are +5 V, will the UJT ever turn on and discharge the capacitor? Why? Can the capacitor ever reach +5 V if $\tau = 0.5T$?

b. Determine and record τ and T for the circuit in Fig. 17-3.

c. How much voltage will be on the capacitor at the end of the first positive pulse? At the end of the second positive pulse? (Assume the voltage does not reach the critical value needed to turn on the UJT.)

d. What value of V_T is required if the capacitor is to discharge at the end of the second positive pulse and the voltage on E of the UJT has to be 60 percent of V_T before the UJT turns on?

e. Construct the circuit in Fig. 17-3 and set V_T to +10 V. Set the generator for 5 $V_{p\text{-}p}$, as indicated on the oscilloscope, and 10 kHz. Have the horizontal controls on the oscilloscope set for 0.2 ms/div. Now view the output on the oscilloscope with the horizontal controls still set for 0.2 ms/div. Describe the output waveform.

f. Now slowly reduce V_T until the oscilloscope displays three or four cycles of the UJT turning on and off. Draw this waveform in Fig. 17-4. What is the voltage of V_T at this time?

g. If the frequency of the square wave is increased to 15 kHz, what should happen to the output waveform? Why? Increase the frequency to 15 kHz and leave it there. Was your answer to the above question correct?

h. Should V_T be increased or decreased so that the UJT can turn on and turn off again? Why?

FIGURE 17-4 Graph for waveform of the RC integrator circuit.

ACTIVITY 17-2 RCL (IMPEDANCE) CIRCUITS

Introduction

This activity provides an opportunity for you to check your understanding of *RCL* circuits. It also provides you with experience in measuring phase shift with an oscilloscope.

If your measured values of I, V, and θ agree with your predicted values within the tolerance of the instruments and components, you have probably mastered *RCL* circuits. If they don't agree within accepted tolerances, try again.

Supplies

- (1) Signal generator, audio range
- (1) Oscilloscope with an ×10 probe
- (1) DMM with current ranges and *f* response to 35 kHz
- (1) Resistor, 330-Ω, ½-W, ±5%
- (1) Resistor, 560-Ω, ½-W, ±5%
- (1) Resistor, 1000-Ω, ½-W, ±5%
- (1) Capacitor 0.01-μF, 25-V, ±10%
- (1) Capacitor 0.05-μF, 50-V, ±10%
- (1) Inductor (RF choke), 2.5 mH, <10-Ω dc resistance, ±10%

Procedure

1. Series *RC* circuit
   a. For the circuit in Fig. 17-5, calculate and record the values needed to complete the first column of the table in Fig. 17-6.
   b. What would happen to θ and V_R if the *f* of this circuit were increased?
   c. What would happen to I_T and V_C if the capacitance were decreased?
   d. Construct the circuit shown in Fig. 17-5, but do not connect the oscilloscope at this time. Set the generator frequency controls for 5 kHz. Then, with the circuit connected to the generator, adjust the generator's output voltage for 5 V as

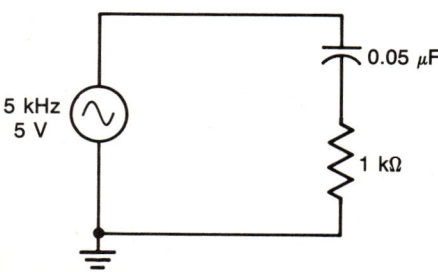

FIGURE 17-5 Series *RC* circuit.

	Series *RC* circuit	
	Calculated	Measured
I_T		
V_R		
V_C		
θ		

FIGURE 17-6 Data table for series *RC* circuit.

measured with the DMM. Using the DMM, measure the other voltages (V_R and V_C) and the circuit current. Record these measured values in Fig. 17-6. Do these measured values agree within 15 percent with the values you calculated in step 1a? If not, check both your calculations and your measurements.

e. θ can be measured with the oscilloscope by using external triggering and observing how far V_R is displaced from V_T. We know that V_R is in-phase with I_T, so the phase angle between V_T and V_R is equal to θ. The displacement between the V_T and V_R waveforms can be determined by following this procedure:

- Connect the ×10 probe to the vertical input of the oscilloscope.
- Set the scope for external triggering and connect a lead from the external trigger input jack to the "hot" output terminal of the signal generator.
- Connect the vertical input of the oscilloscope (through the ×10 probe) to the output of the signal generator. Be sure that the ground lead of the oscilloscope is connected to the ground terminal of the signal generator.
- Vertically center the displayed waveform. Then adjust the horizontal sweep controls and centering control so that one-half cycle spans six divisions on the graticule. This calibrates the *X* axis so that each division is equal to 30° (180°/6 div. = 30°/div.). Accurately observe where the waveform crosses the *X* axis. This is the V_T waveform.
- Now, move the vertical input probe to the top end of the resistor to view the V_R waveform. Observe how far (how many divisions) the new waveform (V_R) is shifted from the old waveform (V_T). Determine θ by multiplying the number of divisions of shift between V_T and V_R by 30°/div.

Record the measured value of θ in Fig. 17-6. Does it agree, within tolerances, with the calculated values? If not, check your measurement and calculation. Did V_R shift left or right from V_T? Does this indicate I_T leads or lags V_T?

f. Increase the generator frequency to 7 kHz. Check the output voltage with the DMM to see if it is still 5 V. Adjust to 5 V if necessary. Now measure V_R and θ. Record them. Was your answer to step 1b correct?

g. Change the capacitor in Fig. 17-6 to 0.01 µF and return the generator's output to 5 kHz, 5 V. Now measure and record I_T and V_C. Was your answer to step 1c correct?

2. Parallel RC circuit
    a. For the circuit shown in Fig. 17-7, calculate the values called for in the table in Fig. 17-8. Record these values in Fig. 17-8, column 1.
    b. Construct the circuit shown in Fig. 17-7. Following the procedures given in step 1d, make the current measurements needed to complete the table in Fig. 17-8. Theta can be measured by inserting a 4.7-Ω resistor between the grounded output lead of the generator and the junction of the 1-kΩ resistor and the capacitor. This added 4.7 Ω of resistance is so small compared to the impedance, resistance, and reactance of the original circuit that its effect on the circuit can be ignored. The few millivolts dropped across the 4.7-Ω resistor will be in phase with I_T, so the technique described in step 1e can be used to measure θ. Record the measured values and compare them to the calculated values. If any values are in disagreement by more than 15 percent, you have probably made an error.
    c. If you increase the frequency in this circuit to 7 kHz, what will happen to I_R and I_T? Make this frequency change; then measure and record I_R and I_T.

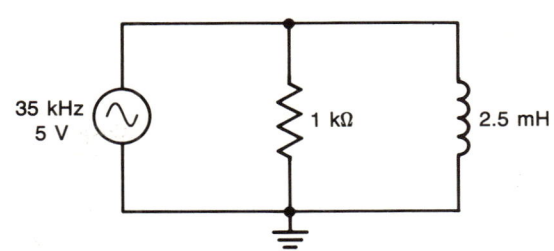

FIGURE 17-9 Parallel RL circuit.

	Parallel RL circuit	
	Calculated	Measured
I_T		
I_R		
I_L		
θ		

FIGURE 17-10 Data table for parallel RL circuit.

3. Parallel RL circuit
    a. For the circuit in Fig. 17-9, calculate and record the values called for in Fig. 17-10.
    b. Construct the circuit and measure (and record) the currents needed to complete the table in Fig. 17-10. Check for unacceptable disagreements between measured and calculated values.
    c. In this circuit, will I_T lead or lag V_T?
    d. Will increasing f cause I_T in this circuit to increase or decrease? Try it.

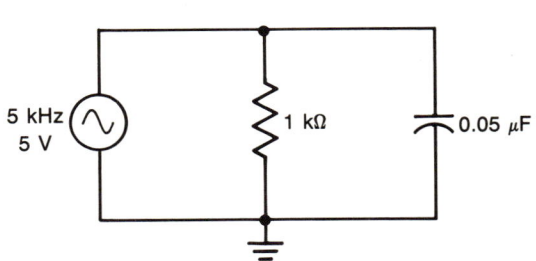

FIGURE 17-7 Parallel RC circuit.

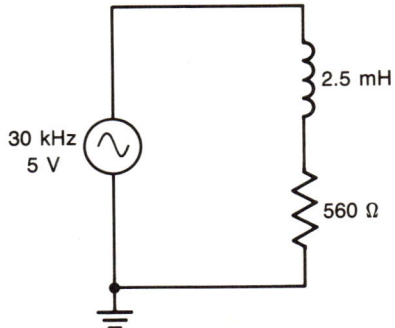

FIGURE 17-11 Series RL circuit.

	Parallel RC circuit	
	Calculated	Measured
I_T		
I_R		
I_C		
θ		

FIGURE 17-8 Data table for parallel RC circuit.

	Series RL circuit	
	Calculated	Measured
I_T		
V_R		
V_L		
θ		

FIGURE 17-12 Data table for series RL circuit.

4. Series *RL* circuit
   a. Refer to Fig. 17-11. Calculate the values needed to complete the calculated column in Fig. 17-12.
   b. Construct the circuit and, using the procedures of steps 1d and e, make the measurements needed to complete Fig. 17-12. Do measured and calculated values agree within expected tolerances?
5. Series *RCL* circuit
   a. Can V_C exceed the value of V_T in a series *RCL* circuit?
   b. If $X_L > X_C$, will I_T lead or lag V_T in a series *RCL* circuit?
   c. For the circuit in Fig. 17-13, record the calculated values called for in Fig. 17-14. Next, construct the circuit and make (and record) the measurements needed to complete the table in Fig. 17-14. If unexplainable disagreements appear in this table, check your work.
   d. Do the data you have collected support your answers to steps 5a and b?
   e. If you change the capacitor in this circuit to 0.05 μF, what will happen to I_T, V_L, and V_C? Change the capacitor and check your answer.
6. Parallel *RCL* circuit
   a. For the circuit in Fig. 17-15, will $I_T < I_R$?
   b. For the circuit in Fig. 17-15, will $I_T < I_C$?
   c. For the circuit in Fig. 17-15, will $I_C < I_L$?
   d. Construct the circuit in Fig. 17-15 and check your answers to a, b, and c above. Record the calculated and measured values of I_T, I_R, I_L, and I_C in Figure 17-16.
   e. If you increase *f* in this circuit to 45 kHz, will I_T increase or decrease? Why? Try it.
7. Review question
   Determine V_T, θ, Z, and V_C for a series *RCL* circuit when $f = 100$ Hz, $R = 100$ Ω, $I_T = 0.01$ A, $L = 2$ H, and $X_C = 796$ Ω.

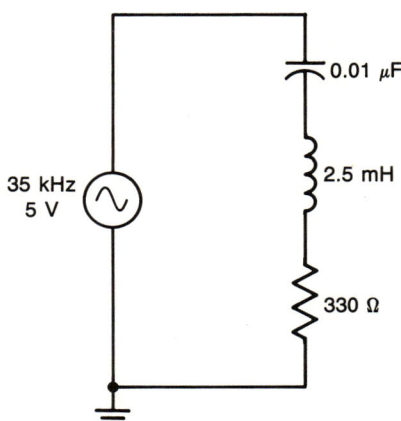

FIGURE 17-13 Series *RCL* circuit.

	Series *RCL* circuit	
	Calculated	Measured
I_T		
V_R		
V_C		
V_L		
θ		

FIGURE 17-14 Data table for series *RCL* circuit.

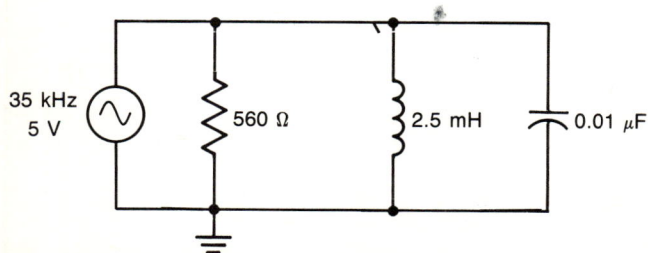

FIGURE 17-15 Parallel *RCL* circuit.

	Parallel *RCL* circuit	
	Calculated	Measured
I_T		
I_R		
I_C		
I_L		

FIGURE 17-16 Data table for parallel *RCL* circuit.

ACTIVITY 17-3 *RC*, *RL*, AND *RCL* CIRCUITS

Introduction

In the previous activity, all component values were given and you determined the results of applying a specified frequency and voltage to the components. In this activity, the procedure is reversed. The desired results are specified, and you are to determine the required components, frequency, measurements, etc.

Supplies

(1) Signal generator, audio range
(1) Oscilloscope with probes
(1) DMM with ac ranges

Miscellaneous resistors, capacitors, and inductors selected from the Materials List in this manual.

Procedure

1. Design, construct, and test a series *RC* circuit that produces 36° of phase shift.
2. Design, construct, and test a parallel circuit in which I_T lags V_T by 72°.
3. Design, construct, and test a circuit in which $V_C = 2V_R$ and $V_L = 2V_C$.
4. Write an activities report which meets or exceeds the specifications given in Appendix 1.

NAME _____ DATE _____

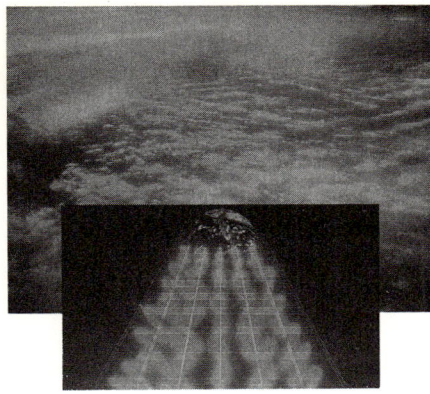

Chapter 18

Resonance and Filters

ACTIVITY 18-1 RESONANT CIRCUITS

Introduction

This activity is concerned with experimentally determining f_r, Q, BW, and resonant voltage rise. The experimentally determined values will then be compared to the theoretically determined values. This way you can check both your theoretical understanding of these concepts and the accuracy of your laboratory techniques.

Supplies

(1) Signal generator, audio range to 500 kHz
(1) Oscilloscope
(1) DMM with f response to 35 kHz
(1) Inductor (RF choke), 2.5-mH, <10-Ω dc resistance, ±10%
(1) Capacitor, 0.01-µF, 25-V dc, ±10%
(1) Capacitor, 0.05-µF, 25-V dc, ±10%
(1) Resistor, 220-Ω, ½-W, ±5%
(1) Resistor, 10-kΩ, ½-W, ±5%

Procedure

1. Series resonance
   a. The resistor in Fig. 18-1 has been included to lower the Q of the circuit so that the Z at resonance will not be too small and overload the signal generator. For the circuit in Fig. 18-1, calculate and record in the table the values called for in Fig. 18-2. In making these calculations, assume an ideal inductor and determine f_{up} and f_{lo} using the formulas $f_{up} \approx f_r + 0.5\ BW$ and $f_{lo} \approx f_r - 0.5\ BW$. The voltages in this table are to be determined at resonance.
   b. What is the resonant voltage rise for this circuit?
   c. The voltage across R is directly proportional to the circuit current. What percentage of the maximum V_r should be across R at f_{up} and f_{lo}?
   d. Construct the circuit in Fig. 18-1. Connect the oscilloscope to the generator output. Use external triggering and have the vertical controls in their calibrated position. Set the generator to the resonant frequency you calculated in step 1(a). Adjust the output for 5 V (14 V_{p-p}). Connect the

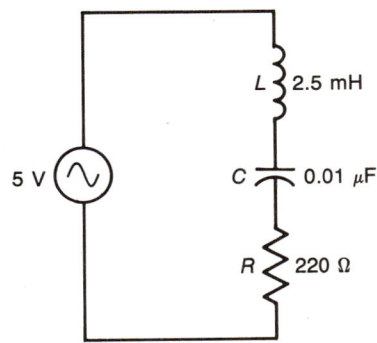

FIGURE 18-1 Series resonant *RCL* circuit.

	Series *RCL*	
	Calculated	Measured
f_r		
Q		✕
BW		
V_C		
V_L		
V_R		
f_{lo}		
f_{up}		

FIGURE 18-2 Data table for series resonant *RCL* circuit.

Copyright © 1993 by the Glencoe Division of Macmillan/McGraw-Hill School Publishing Company. All rights reserved.

DMM across the resistor. Should the meter be indicating minimum or maximum voltage at this frequency? Slowly adjust the generator frequency until maximum voltage appears across the resistor. Check input voltage and readjust if necessary to be sure it remains at 14 $V_{p\text{-}p}$. Record the measured values of V_R and f_r in Fig. 18-2. Now measure V_L and V_C with the DMM; record their values.

Using the techniques used in the previous activity, measure and record the phase angle between V_R and V_T. Is this angle 0?

e. Decrease the generator frequency until V_R decreases to 70.7 percent of its maximum value. Again check to be sure the source voltage is still 14 $V_{p\text{-}p}$ and adjust the source voltage if necessary. Record this frequency in Fig. 18-2. Besides f_{lo}, what other terms could be used to describe this frequency? Should V_R be leading or lagging V_T at this time? By how many degrees? Check your answers by measuring the angle with the oscilloscope. Were you correct?

f. Using the procedure described in step e above, determine and record f_{up}. Is V_R leading or lagging V_T at this time?

g. Measure and record the ohmic resistance of the 2.5-mH inductor you have been using. Could this resistance account for some of the differences between measured and calculated values in Fig. 18-2?

h. Using the measured value of V_C and V_T, calculate Q. Would you expect this value to be within 20 percent of the previously calculated value? Why?

i. Within expected tolerances, does the measured f_r divided by the measured BW equal Q? Should it?

j. If you change C in Fig. 18-1 to 0.05 μF, what should happen to f_r, V_C, and Q? Check your answers by changing C and making the necessary adjustments and measurements. Record f_r and V_C.

2. Parallel resonance
   a. Should the resonant frequency for the circuit in Fig. 18-3 be approximately the same value as for the circuit in Fig. 18-1? Why?
   b. What relationship should exist between I_T, Q, and I_C for the circuit in Fig. 18-3?
   c. Calculate and record in Fig. 18-4 the values needed for the first column of the table in Fig. 18-4. Use the same formulas for calculating f_{up} and f_{lo} as were used for the series-resonant circuit. Again, assume ideal components.
   d. Construct the circuit in Fig. 18-3 and again use the oscilloscope to monitor the generator output voltage. Be sure the output voltage is 5 V (14 $V_{p\text{-}p}$) before taking and recording any cur-

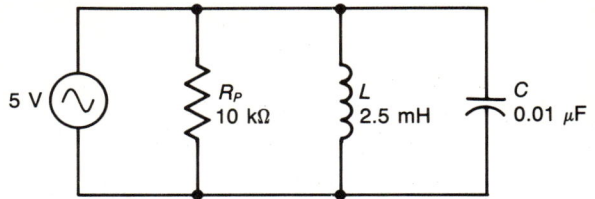

FIGURE 18-3 Parallel resonant RCL circuit.

	Parallel RCL	
	Calculated	Measured
f_r		
Q		✗
BW		
I_C		
I_L		
I_R		
I_T		
f_{lo}		
f_{up}		

FIGURE 18-4 Data table for parallel resonant RCL circuit.

rent measurements. Using the procedures outlined for the series-resonant RCL circuit, make the measurements needed to complete the table in Fig. 18-4. Remember that f_{up} and f_{lo} are the frequencies at which Z is 0.707 of its maximum value and that I_T will be 1.414 times its minimum value when Z is 0.707 of its maximum value.

e. What should happen to I_T in a resonant parallel RCL circuit when R_P is removed? Check your answer by removing R_P in Fig. 18-3 and again measuring I_T.

ACTIVITY 18-2 FILTERS

Introduction

This activity investigates the characteristics of both resonant and nonresonant filters. It demonstrates how a filter separates two frequencies. In the final part of the activity, a double resonant circuit is constructed and tested.

Supplies

(1) Signal generator, audio range to 500 kHz
(1) Oscilloscope with ×10 probe

NAME _____ DATE _____

(1) DMM with frequency response to 35 kHz
(1) Inductor (RF choke), 1.0-mH <10-Ω dc resistance, ±10%
(1) Inductor (RF choke), 2.5-mH <10-Ω dc resistance, ±10%
(1) Capacitor, 0.01-μF, 25-V dc, ±10%
(1) Resistor, 100-Ω, ½-W, ±5%
(1) Resistor, 1-kΩ, ½-W, ±5%

Procedure

1. Nonresonant filter
  a. Is the circuit in Fig. 18-5 a low-pass or high-pass filter?
  b. What is the cutoff frequency for this filter?
  c. What is the −20-dB frequency for this filter?
  d. What should be the value of V_R at −20 dB?
  e. What is the value of V_R at −14 dB?
  f. At what frequency should the output be 45° out-of-phase with the input? Will the output lead or lag the input?
  g. Construct the circuit in Fig. 18-5 and connect the oscilloscope across the generator output. Set the vertical controls on the oscilloscope in their calibrate position. Adjust the generator output voltage for 10 $V_{p\text{-}p}$ each time the generator frequency is changed. For each frequency listed in Fig. 18-6, measure the output voltage across the resistor and record the peak-to-peak value in the table. If the DMM you are using is accurate at these frequencies, use it to measure V_R. If not, first use the oscilloscope to set the input voltage at 10 $V_{p\text{-}p}$ and then to measure the output voltage.
  h. Now convert each of the voltages in Fig. 18-6 to decibels and plot a decibel output versus frequency curve on the semilog paper in Fig. 18-7. Were your answers to the questions in steps 1b, c, d, and e correct within the expected tolerances? If not, repeat your measurements and/or calculations.
  i. Now that we know the characteristics of this filter, let us use it to improve the signal-to-noise

f (kHz)	V_o ($V_{p\text{-}p}$)
1	
2	
3	
4	
5	
6	
7	
8	
9	
10	
20	
30	
40	
50	
60	
70	
80	
90	

FIGURE 18-6 Data table for *RL* filter circuit.

ratio of a complex signal. Figure 18-8 shows a circuit which combines the output of two generators to provide a 90-kHz, 0.5-$V_{p\text{-}p}$ waveform riding (or superimposed) on a 3-kHz, 5-$V_{p\text{-}p}$ signal. This complex signal is then applied to the low-pass filter for which you just collected data. If the 3-kHz waveform is the desired signal, what is the S/N ratio of the signal applied to the filter in Fig. 18-8?
  j. Using the data collected in Fig. 18-6, estimate and record the expected output voltage at 3 and 90 kHz.
  k. What is the expected S/N ratio of the signal out of the filter? (Use the voltages estimated in step 1j.)
  l. Construct the circuit in Fig. 18-8 and connect the oscilloscope across the input of the filter. Adjust the output voltage of the 90-kHz generator for 0.5 $V_{p\text{-}p}$. Next, adjust the output voltage of the 3-kHz generator until the peak-to-peak value of the complex signal is 5.5 V. (When the complex signal is 5.5 $V_{p\text{-}p}$, the 3-kHz generator output will be 5 $V_{p\text{-}p}$.) Now move the oscilloscope to the output terminals of the filter and view the output waveform. Although it is difficult to measure the exact voltage of the 90-kHz wave-

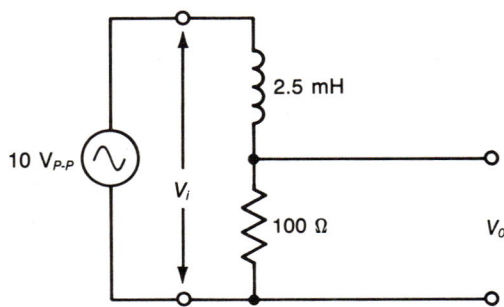

FIGURE 18-5 *RL* filter circuit.

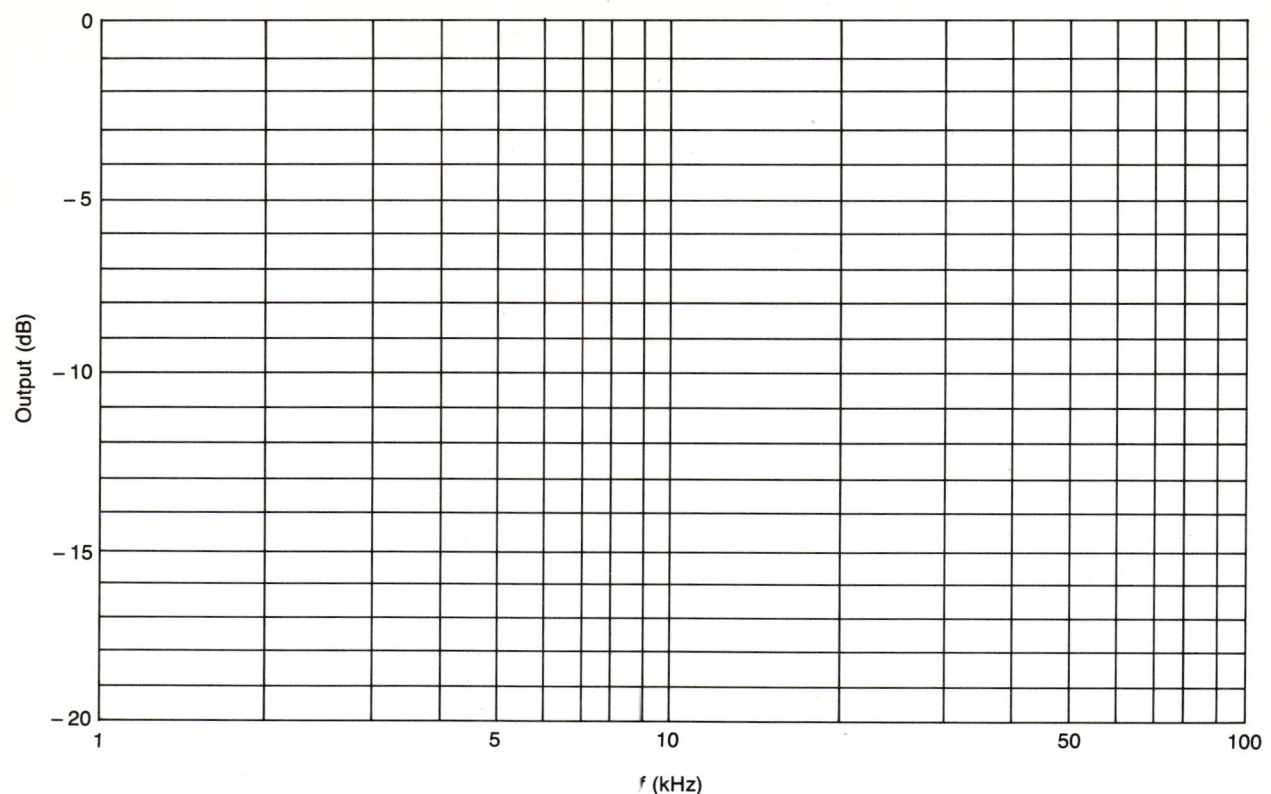

FIGURE 18-7 Frequency response of the *RL* filter.

form, does it appear to have been attenuated to the expected level? What is the voltage level of the 3-kHz signal at the output of the filter? Within expected tolerances, is this the level predicted?

2. Double-resonant filter
   a. Figure 18-9 shows a double-resonant filter. Using the values shown, calculate (and record) the band-reject frequency.
   b. Now calculate (and record) the value of L_2 needed to provide a bandpass frequency of 59.6 kHz.
   c. Construct the circuit in Fig. 18-9 using the value for L_2 just calculated. Set the generator for

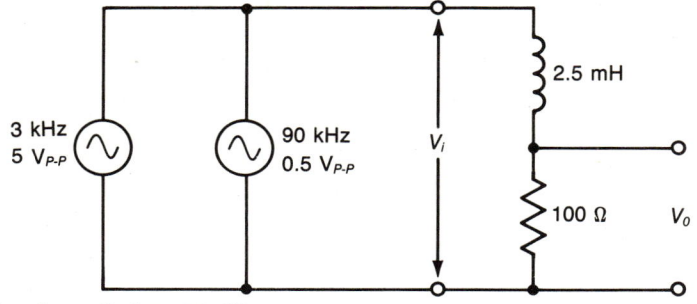

FIGURE 18-8 Complex signal applied to *RL* filter.

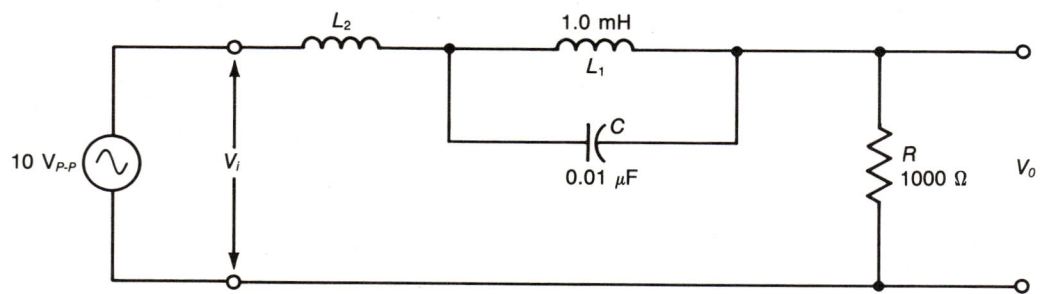

FIGURE 18-9 Double-resonant-filter circuit.

f (kHz)	V_o (V$_{p\text{-}p}$)
10	
20	
30	
40	
50	
60	
70	
80	
90	
100	
200	
300	
400	
500	
600	
700	
800	
900	

FIGURE 18-10 Data table for double-resonant filter.

59.6 kHz and adjust the output for 10 V$_{p\text{-}p}$ as indicated on the oscilloscope. Now, while monitoring the output of the filter with the oscilloscope, slowly change the generator frequency around 59.6 kHz to obtain the highest possible output. Check to be sure that V_i is still 10 V$_{p\text{-}p}$. Record both the frequency and the voltage when the output is maximum. If the frequency isn't between 55 and 65 kHz, check your circuit and/or calculations.

d. Change the generator frequency to that calculated in step 2a. Keep the input voltage at 10 V$_{p\text{-}p}$. This time, slowly rock the generator frequency control to obtain the lowest possible output voltage when V_i is 10 V$_{p\text{-}p}$. Record the frequency and voltage when the output is minimum. Does the frequency agree, within tolerances, with the frequency calculated in step 2a? If not, check your work.

e. While maintaining V_i at 10 V$_{p\text{-}p}$, measure (and record) V_o for each frequency listed in Fig. 18-10.

f. Using the data collected in steps 2c, d, and e, draw an f versus V_o curve on the semilog paper in Fig. 18-11.

g. From the curve just completed, estimate the three -3-dB frequencies for this double-resonant circuit. (Remember that -3-dB frequencies occur when the output is 70.7 percent of maximum.)

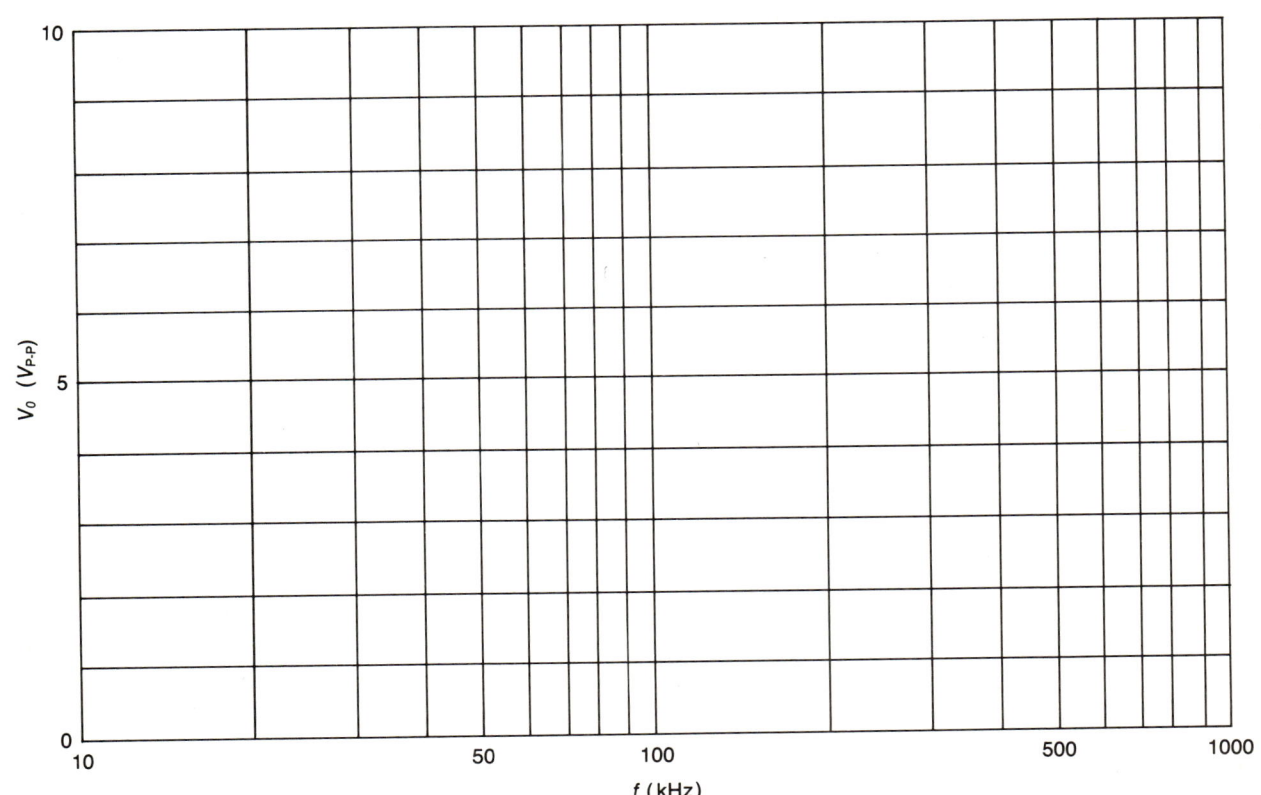

FIGURE 18-11 Frequency response of double-resonant filter.

ACTIVITY 18-3 DESIGNING FILTERS

Introduction

Filters are extensively used in electronic circuits. Designing simple filters to produce specified results, rather than analyzing existing filters, helps one to understand this important type of circuit.

Supplies

(1) Signal generator, audio range to 100 kHz
(1) DMM with ac ranges
(1) Oscilloscope with ×10 probe

Miscellaneous resistors, capacitors, and inductors selected from the Materials List in this manual.

Procedure

1. Assuming ideal inductors (i.e., $X_L \gg R_L$), design, construct, and test a parallel *RCL* circuit that is resonant at 22.5 kHz and has a bandwidth of 6 kHz.

2. Using a 10-kΩ resistor as one element of the filter, design, construct, and test a high-pass filter with a −10-dB frequency of 1060 Hz and an output voltage that leads the input voltage.

3. Using a 2.5-mH inductor (assume that it is ideal), design, construct, and test a band-pass filter that has an upper half-power point of 52 kHz and a bandwidth of 14 kHz.

4. Produce a written report that meets or exceeds the requirements specified in Appendix 1.

NAME _____ DATE _____

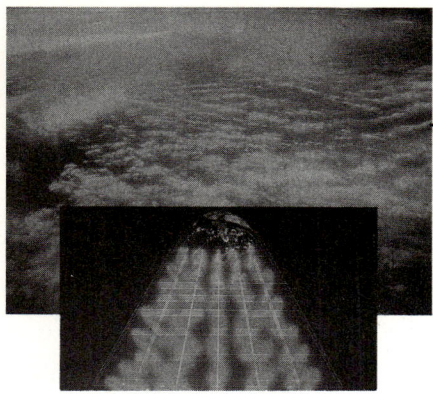

Chapter 19

AC Network Analysis

ACTIVITY 19-1 SERIES-PARALLEL AC CIRCUITS

Introduction

This activity provides experimental verification of the theoretical analysis of a complex ac circuit. After using phasor algebra to analyze a series-parallel RC circuit to determine the currents and voltages, the circuit will be constructed and the magnitude of all currents and voltages will be measured. Also, the phase angle of one voltage will be measured to verify phase relationships as well as magnitude relationships.

Supplies

(1) Signal generator, audio range
(1) DMM, 10-MΩ input resistance
(1) Oscilloscope with $\times 10$ probe
(1) Resistor, 1-kΩ, $\frac{1}{4}$-W, $\pm 5\%$
(2) Resistors, 2.2-kΩ, $\frac{1}{4}$-W, $\pm 5\%$
(2) Capacitors, 0.1-μF, $\pm 5\%$, 25 DCWV

Procedure

1. **a.** The voltages and currents in Fig. 19-1 can be calculated by first determining the impedance (both magnitude and angle) of C_1 in series with R_1 and C_2 in parallel with R_2. These two impedances are then added, using vector algebra, to find Z_T. With Z_T and V_T, the total current (and thus I_{R_1}, I_{C_1}, and I_{Z_2}) can be determined and used to calculate individual voltages. Once the voltage across Z_2 (the parallel combination of R_2 and C_2) is known, I_{C_2} and I_{R_2} can be calculated. Using this procedure make and record the calculations specified in Fig. 19-2.
   b. Check your calculations in step 1a above by seeing if $V_{R_2} = V_T - V_{R_1} - V_{C_1}$. Does it? If not, recheck your calculations.

2. **a.** Construct the circuit in Fig. 19-1. Be sure the generator's output voltage is adjusted for 5 V, as indicated by the DMM, *after* the generator is connected to the circuit. (Because of the generator's significant internal impedance, its output voltage will vary as the load on it changes.) Using the DMM, measure (and record) the voltages and currents called for in the table in Fig. 19-2.
   b. Do the magnitudes of the measured and calculated values agree within ± 10 percent? If not, recheck your measurements and/or calculations.

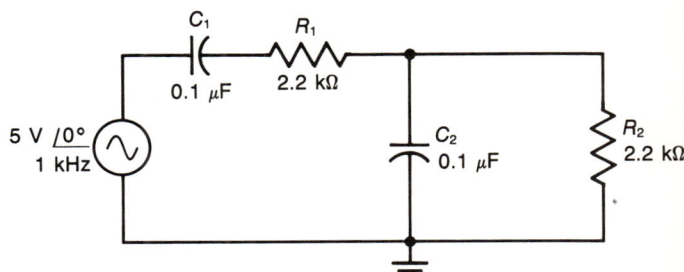

FIGURE 19-1 Series-parallel RC circuit.

	Calculated		Measured
	Magnitude	Angle	Magnitude
I_T			
I_{R_2}			
I_{C_2}			
V_{R_1}			
V_{R_2}			
V_{C_1}			

FIGURE 19-2 Data table for the circuit in Fig. 19-1.

Copyright © 1993 by the Glencoe Division of Macmillan/McGraw-Hill School Publishing Company. All rights reserved.

83

3. **a.** Refer to Fig. 19-1 and notice that V_T is specified as 5 V $\underline{/0°}$. This means that V_T would be on the reference axis in a phasor diagram for the voltages and currents in this circuit. Now, from your calculated data in Fig. 19-2, determine and record the phase angle between V_T and V_{R_2}. Is V_{R_2} leading or lagging V_T?

b. Using the oscilloscope, with the ×10 probe, measure the phase angle between V_{R_2} and V_T. The procedure for doing this is:

- Set the scope for external triggering and connect a lead from the external trigger input jack to the "hot" output terminal of the signal generator.
- Connect the vertical input of the oscilloscope to the output of the signal generator. Be sure that the ground lead of the oscilloscope is connected to the ground terminal of the signal generator.
- Vertically center the displayed waveform. Then adjust the horizontal sweep controls and centering control so that one-half cycle spans six divisions on the graticule. This calibrates the X axis so that each division is equal to 30° (180°/6 div. = 30°/div.). Accurately observe where the waveform crosses the X axis.
- Move the vertical input probe to the top end of R_2 (this measures V_{R_2}) and observe how far the new waveform (V_{R_2}) is shifted from the old waveform (V_T). Calculate and record the phase angle between V_T and V_{R_2}.

c. Is V_{R_2} shifted to the left or the right from V_T? Does this indicate V_{R_2} is leading or lagging V_T?

ACTIVITY 19-2   NORTON AND THEVENIN EQUIVALENT AC CIRCUITS

Introduction

In this activity you will predict V_{Th} and I_N for a series-parallel ac circuit; then, you will measure V_{Th} and I_N to check your prediction. Next you will predict, and then measure, the terminal voltage of the circuit when a load is applied. Finally, after determining the RC values needed to provide Z_{Th}, you will construct and test the Thevenin equivalent circuit to determine if it behaves as the original circuit behaved.

Supplies

(1) Signal generator, audio range
(1) DMM, 10-MΩ input resistance
(1) Oscilloscope with ×10 probe
(1) Resistor, 1-kΩ, ¼-W, ±5%
(1) Resistor, 2.2-kΩ, ¼-W, ±5%
(2) Capacitors, 0.1-μF, ±5%, 25 DCWV

Procedure

1. **a.** Figure 19-3 shows an ac circuit to which a load can be connected between terminals A and B. The circuit can be converted to its Thevenin equivalent by calculating V_{Th} and Z_{Th}. V_{Th} is found by determining the voltage (both magnitude and angle) between A and B when no load is attached. Z_{Th} is found by determining the impedance (magnitude and angle) between A and B when the generator is replaced by a conductor. (This assumes the internal impedance of the generator is 0 Ω; i.e., its output voltage remains at 5 V even though the load current is changed). Calculate Z_{Th} and V_{Th} for Fig. 19-3. Record these calculated values in the table in Fig. 19-4.

   b. Use the values of V_{Th} and Z_{Th} just calculated to calculate I_N for the equivalent Norton circuit. Record the value of I_N in Fig. 19-4.

2. **a.** Construct the circuit in Fig. 19-3. Now, using the DMM, measure the magnitude of the voltage between terminals A and B. Does the measured voltage agree, within 10 percent, with the value you calculated for V_{Th}? Record this value in the table in Fig. 19-4. Using the procedure outlined in the previous activity on series-parallel circuits, measure the phase angle of V_{Th}. Record this angle. Does this angle agree with the calculated angle?

   b. Next, measure the value of I_N for the circuit in Fig. 19-3. Remember that Z_{Th}, and thus I_N, was calculated assuming that the generator had zero internal impedance. Of course, it doesn't. However, you can make the generator appear to have no internal impedance if you adjust its output voltage for exactly 5 V each time the load on the generator is changed.

   To measure I_N, first short terminals A and B together with a jumper wire and adjust the generator's output voltage to 5 V. Remove the jumper wire from between terminals A and B and place the DMM, on the 20-mA range, between terminals A and B. The current measured by the DMM is I_N. Record this value in the table in Fig. 19-4. Does the measured value of I_N agree, within 10 percent, with the calculated value? If not, recheck your circuit and your measurements.

3. **a.** Using your calculated values of Z_{Th} and V_{Th}, predict the voltage between terminals A and B

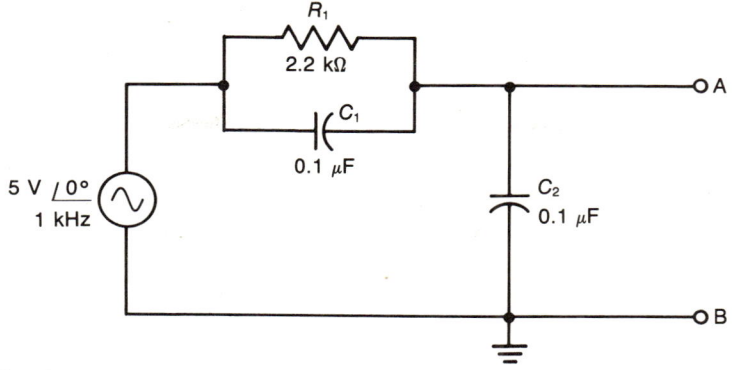

FIGURE 19-3 Circuit to be theveninized.

	Calculated		Measured
	Magnitude	Angle	Magnitude
V_{Th}			
Z_{Th}			
I_N			

FIGURE 19-4 Data table for the circuit in Fig. 19-3.

when a 1-kΩ resistor is connected between these terminals. Record this voltage as V_{RL}.

b. Connect a 1-kΩ resistor between terminals A and B. Adjust the generator's output for 5 V. Now, measure the magnitude of V_{RL} with the DMM and then measure the phase angle with the oscilloscope. Record these values. Do they agree, within tolerances, with the calculated values?

4. a. Now let's replace the circuit of Fig. 19-3 with its Thevenin equivalent circuit by determining what parallel values of R and C will provide Z_{Th}. Of course, the signal generator's output voltage can be adjusted to provide V_{Th}.

Figure 19-5 illustrates the procedure for finding the parallel RC values for a Z_{Th} at $-60°$. The process involves:

- Converting impedance to admittance.
- Splitting admittance into its real and imaginary components: conductance and susceptance.
- Converting conductance and susceptance to resistance and capacitive reactance.
- Calculating the value of C needed to provide X_C at the operating frequency.

All the necessary formulas are given in Fig. 19-5.

Calculate the values of R and C needed to provide the Z_{Th} you recorded in Fig. 19-4. Record these values.

b. Using the values of R and C calculated above, construct the Thevenin equivalent circuit which represents the circuit in Fig. 19-3. Load the Thevenin circuit with a 1-kΩ resistor. Set the output of the signal generator so that its voltage magnitude is equal to the magnitude of V_{Th}. Now, measure, with the DMM, the magnitude of V_{RL}. Record this value. Does this value agree, within expected tolerances, with the value of V_{RL} for the original circuit in Fig. 19-3 (see step 3b)?

c. Finally, measure the phase angle between V_{Th} (the generator's output) and V_{RL}. Record this value. Is V_{RL} leading or lagging V_{Th}? Considering that $V_{Th} = 2.9$ V $\underline{/-16°}$ and $V_{RL} = 2.0$ V $\underline{/13.3°}$, does your measured angle seem to be correct?

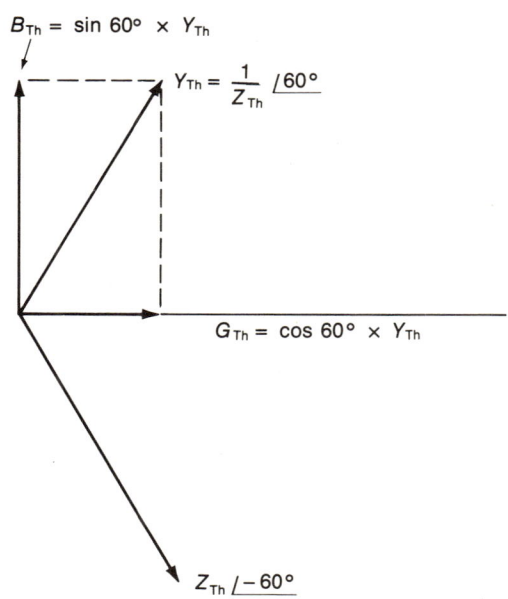

FIGURE 19-5 Converting Z_{Th} into its equivalent parallel RC values.

ACTIVITY 19-3 AC NETWORK PROBLEM

Introduction

In this activity you will determine the components needed to provide specified network impedances. Then, you will determine how the network responds to various loads. Finally, you will write a report detailing how you analyzed and tested the network.

Supplies

(1) Signal generator, audio range to 50 kHz
(1) DMM with ac ranges
(1) Oscilloscope with ×10 probe

Miscellaneous resistors, capacitors, and inductors selected from the Materials List in this manual.

Procedure

1. For the network shown in Fig. 19-5, determine the load voltage and phase angle when the load connected between points A and B is:
   a. $1000\ \Omega\ /\ 0°$
   b. $398\ \Omega\ /\ -90°$
   c. $517\ \Omega\ /\ -50.3°$
2. Determine the components needed to (a) construct the network shown in Fig. 19-6, and (b) provide the specified loads.
3. Construct and test the network with the various loads connected.
4. Write a report that fulfills or exceeds the requirements listed in Appendix 1.

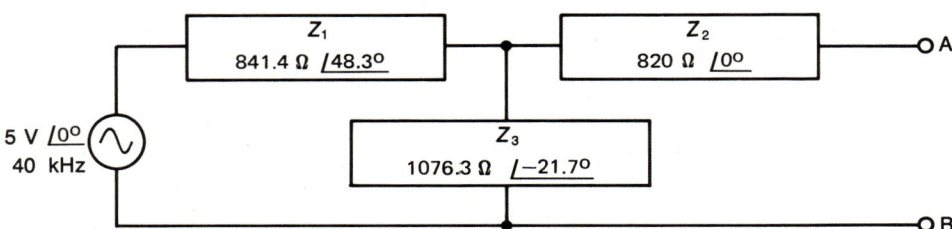

FIGURE 19-6   AC network to be analyzed, constructed, and tested.

NAME _____ DATE _____

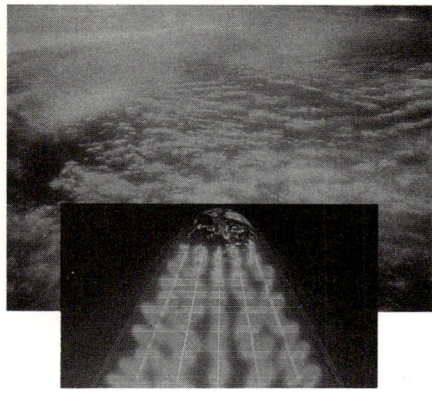

Chapter 20

Transformers

ACTIVITY 20-1   TRANSFORMER VOLTAGE RATIOS AND PHASING

Introduction

One of the purposes of this activity is to investigate the voltage ratios between the windings of a multiprimary, multisecondary transformer. The other purpose is to learn to properly phase windings for a parallel, series-aiding, and series-opposing connection.

Supplies

(1) Transformer, dual 115-V primaries, dual 12-V, 2-A secondaries (Triad F107Z or equivalent)
(1) Fuse, 1-A, 3-AGC with holder
(1) VOM or DMM
(1) Oscilloscope with external triggering and ×1 probe

Procedure

↪ **CAUTION:** The primary voltage used in this activity can cause a fatal shock. Use safe construction practices in connecting this circuit. Do not touch any part of the circuit after power is applied.

1. Voltage ratios
   a. Construct the circuit in Fig. 20-1 and apply power. Measure, and record in the table in Fig. 20-2, the voltages for all four windings. Remove power from the primary.
   b. Are the voltages you recorded in Fig. 20-2 greater than the rated values given on the transformer? Why?
   c. From the data in Fig. 20-2, determine the turns ratio for this transformer when winding 5-7 is used as the secondary.

2. Phasing of windings
   a. Refer to Fig. 20-1. Connect terminal 6 to terminal 7. Apply power. Measure and record the voltage between terminals 5 and 8. Are the two windings connected series-aiding or series-opposing? Remove the primary power.
   b. Connect the oscilloscope to the circuit as shown in Fig. 20-3(a). This arrangement will allow you to compare the phase of the voltage at terminals 5 and 8 when both voltages are referenced to terminal 6 or 7. Start with the vertical input connected to terminal 5. Apply power and adjust the oscilloscope controls to display one cycle which is vertically and horizontally centered. Be sure the oscilloscope is set for external triggering. Draw the observed waveform on the graph in Fig. 20-4. Now move the vertical input to terminal 8 but do not change any other oscilloscope controls or leads. Draw the waveform observed on the same graph in Fig. 20-4. Remove power. What is the phase relationship between the voltages at terminals 5 and 8? Do the observed waveforms explain why the voltage between ter-

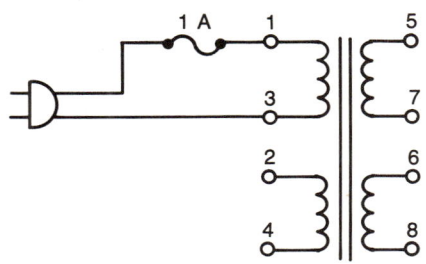

FIGURE 20-1   Circuit for Activity 20-1, step 1.

	Winding between terminals			
	1-3	2-4	5-7	6-8
Voltage				

FIGURE 20-2   Data table for Activity 20-1, step 1a.

Copyright © 1993 by the Glencoe Division of Macmillan/McGraw-Hill School Publishing Company. All rights reserved.

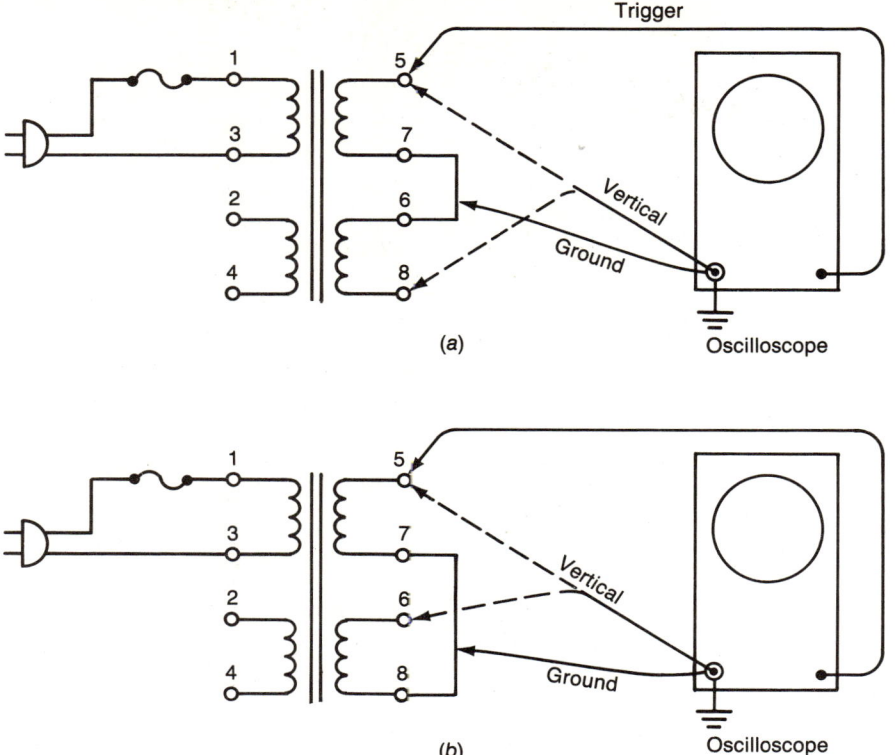

FIGURE 20-3 Circuits for Activity 20-1, steps 2b and d.

minals 5 and 8 is twice as large as the voltage between terminals 5 and 7 or terminals 8 and 6?

c. Refer to Fig. 20-1. If terminals 7 and 8 were connected together, would the secondaries be aiding or opposing? Would the voltages at terminals 5 and 6 be in phase or out-of-phase?

d. Construct the circuit shown in Fig. 20-3(b) and apply power. Using the VOM or DMM, measure and record the voltage between terminals 5 and 6. Using the oscilloscope, determine and record the phase relationship of the voltages at terminals 5 and 6. Remove power. Do your answers in step c agree with your answers in step (d)?

e. Construct the circuit shown in Fig. 20-5. The primaries are connected series-aiding in this circuit so they could be connected to a 230-V line. However, as shown in Fig. 20-5, you will continue to use 115 V (or 120 V). How does the turns-per-volt ratio of this circuit compare to that in Fig. 20-1? How much voltage should be provided by winding 5-7? Apply power and measure and record the voltage between terminals 5 and 7. Remove power. Does the measured voltage support your previous answers?

f. In Fig. 20-6 one of the windings which the manufacturer intended for a secondary is connected in series with the two primaries. If this winding is series-aiding, should the voltage between terminals 5 and 7 be increased or decreased from its value in the previous step? Why? Construct the circuit of Fig. 20-6. Apply power and measure and record the voltage between terminals 5 and 7. Remove power. Is winding 6-8 series-

FIGURE 20-4 Graph for waveforms.

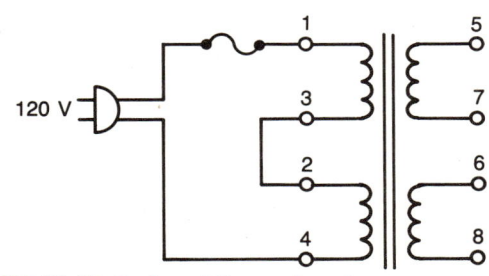

FIGURE 20-5 Series-aiding primaries.

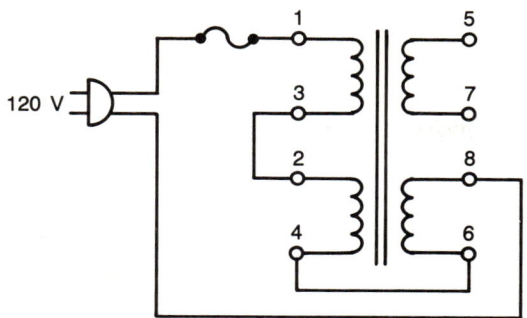

FIGURE 20-6 Circuit for Activity 20-1, step 2f.

aiding or series-opposing the rest of the primary?

g. Should the voltage between terminals 2 and 8 be more than or less than the voltage between terminals 1 and 3? Why?

ACTIVITY 20-2 TRANSFORMER TESTS AND EQUIVALENT CIRCUITS

Introduction

In this activity you will perform an open-circuit and a short-circuit test on a small transformer. Using the test data, you will develop an equivalent circuit for the transformer and predict its performance under load. Finally, you will load the transformer and check your predictions.

Supplies

(1) Variable ac supply, 60-Hz, 0- to 120-V, 1-A
(1) Wattmeter, 75-W, 150-V max., 1-A max.
(1) VOM or DMM with ac ranges
(1) Resistor, 12-Ω, 100-W, ±5%
(1) Transformer, dual 115-V primaries, dual 12-V, 2-A secondaries (Triad F107Z or equivalent)
(1) Fuse, 1-A, 3-AGC with holder

Procedure

CAUTION: The primary voltage used in this activity can cause a fatal shock. Use safe construction practices in connecting this circuit. Do not touch any part of the circuit after power is applied.

1. Open-circuit test (before connecting to the ac supply)
   a. Measure, and record in Fig. 20-7, the ohmic resistance of the transformer primary winding and secondary winding when connected as shown in Fig. 20-8.
   b. Construct the circuit shown in Fig. 20-8 and set the autotransformer for zero output. Apply power and adjust the autotransformer until the voltmeter indicates 115 V. Read (and record in Fig. 20-9, row 1) the indicated power, voltage, and current for the primary. Also record the indicated current in Fig. 20-7. Turn off the power but do not change the autotransformer setting. If another voltmeter is not available, remove the voltmeter from the primary. Then apply power and measure the open-circuit secondary voltage. Record this value for use later on. Remove power and reconnect the voltmeter in the primary if it was removed.
   c. The open-circuit test is designed to measure the core loss of a transformer, but the wattmeter also indicates the power lost in the primary

Winding	Resistance	Open-circuit		Short-circuit	
		Current	Power loss	Current	Power loss
Primary					
Secondary					

FIGURE 20-7 Winding data.

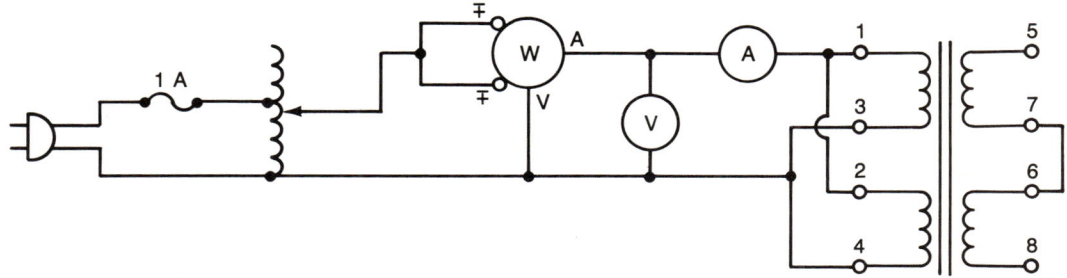

FIGURE 20-8 Circuit for transformer tests.

	Primary current (I_p)	Primary voltage (V_p)	Primary power (P_p)
Open-circuit (OC)			
Short-circuit (SC)			

FIGURE 20-9 Power losses.

windings. Using the data in Fig. 20-7 and $P = I^2R$, determine the power used by the primary windings. Record this power in row 1, column 3 of Fig. 20-7. What percentage of the indicated power was lost in the windings?

2. Short-circuit test
    a. To perform the short-circuit test one needs to know the full-load primary current. This can be determined by applying the full load to the secondary. Since the secondary is rated at 24 V, 2 A when the two windings are series-aiding, a 12-Ω resistor (R_L) will provide a full load. Connect a 12-Ω, 100-W resistor between terminals 5 and 8 of Fig. 20-8. *This resistor will get very hot when power is applied to the circuit.* Apply power; adjust the autotransformer for a voltmeter indication of 115 V; read the indicated current and power; turn off the power, but do not change the autotransformer setting. Record the indicated current and power in Fig. 20-10. If you do not have another ac voltmeter, remove the voltmeter from the primary and connect it across the load resistor. Apply power and measure the voltage across the load resistor. Remove power. Record the load voltage in Fig. 20-10. Disconnect the 12-Ω load resistor; measure its value; record this value in Fig. 20-10. If the voltmeter was removed from the primary circuit, replace it.
    b. Connect a low-resistance jumper wire between terminals 5 and 8 in Fig. 20-8. *Adjust the autotransformer for zero output* and then apply power to the circuit. Now, slowly increase the output of the autotransformer until the ammeter indicates the full-load primary current determined in step 2a above. Read (and record in Fig. 20-9, row 2) the indicated power, current, and voltage for the primary. Remove power.
    c. The power measured by the short-circuit test is essentially the power lost in the windings. You can calculate the power lost in the windings and compare it with the power measured by the wattmeter. First, record the short-circuit primary and secondary currents in column 4 of Fig. 20-7. (The short-circuit secondary current will be 2 A because the short-circuit primary current was adjusted to the full-load value.) Now, using $P = I^2R$, complete the last column of Fig. 20-7 and add the values in the column to determine the power used by the windings. Record this value. What percentage of the power indicated by the wattmeter is this calculated power?
    d. Other than meter and meter-reading errors, what accounts for the difference between the power indicated by the wattmeter in step 2b and the winding power calculated by I^2R in step 2c?

3. θ and efficiency
    a. Using the load voltage and load resistance recorded in Fig. 20-10, calculate and record the power delivered by the secondary to the load.
    b. Now, using the data in Fig. 20-10 (P_p and P_{RL}), calculate and record the efficiency of the transformer under full-load conditions.
    c. The efficiency can also be determined by

    $$\eta = \frac{P_{\text{out}}}{P_{\text{out}} + P_{\text{copper loss}} + P_{\text{core loss}}}$$

    Using the appropriate data from Fig. 20-9 and Fig. 20-10, calculate the efficiency using this formula. In addition to meter error and meter-reading error, why might there be a difference in the efficiencies calculated by the two formulas?
    d. Should θ (the angle between the primary current and the primary voltage) be largest under no-load or full-load conditions?
    e. Check your answer to step 3d above by calculating the two θ's with the data available in Fig. 10-

	Primary current I_p	Primary voltage V_p	Primary power P_p	Angle theta θ	Load voltage V_{R_L}	Load resistance R_L	Load power P_{R_L}	% of efficiency
No-load					✕	✕	✕	✕
Full-load								

FIGURE 20-10 Angle θ and efficiency.

10. Record your answers in this figure. The appropriate formula can be derived from $P = IV \cos\theta$

$$\theta = \arccos\frac{P}{IV}$$

4. Equivalent circuits
   a. The data from the open-circuit test and the closed-circuit test can be used to estimate the constants needed for the equivalent circuits shown in Fig. 20-11(a) and (b). From the open-circuit primary and secondary voltages found in step 1a, determine and record the transformation ratio for the transformer with the two windings connected series-aiding to form a single secondary.
   b. Using formulas below and data from Fig. 20-9, calculate and record the values specified in Fig. 20-11(c).

$$R_{CL} = \frac{V_{P_{OC}}^2}{P_{OC}}$$

$$Z_{OC} = \frac{V_{P_{OC}}}{I_{P_{OC}}}$$

$$X_M = \sqrt{\frac{Z_{OC}^2 \, R_{CL}^2}{R_{CL}^2 - Z_{OC}^2}}$$

$$R_E = \frac{P_{SC}}{I_{P_{SC}}^2}$$

$$Z_{SC} = \frac{V_{P_{SC}}}{I_{P_{SC}}}$$

$$X_E = \sqrt{Z_{SC}^2 - R_E^2}$$

   c. According to the equivalent circuits, $R_E = R_P + a^2 R_S$. Using the values of R_P and R_S recorded in Fig. 20-7 and the value of a determined in step 4a, calculate the value of R_E with the above formula. Record this value. How does it compare with the value recorded in Fig. 20-11(c)?
   d. Using the value of R_L from Fig. 20-10, determine and record the value of $a^2 R_L$.
   e. You now have all the values needed to determine the load power by analyzing the equivalent circuit of Fig. 20-11(b). Do so, and record the load power. The necessary formulas are

$$R_T = R_E + a^2 R_L$$
$$Z = \sqrt{R_T^2 + X_E^2}$$
$$I = \frac{115 \text{ V}}{Z}$$
$$P_{R_L} = I^2 \times a^2 R_L$$

   f. The equivalent circuit of Fig. 20-11(b) can also be used to estimate the efficiency. The power used by R_E is an estimate of the transformer losses (note that this simplified equivalent ignores R_{CL} and its power loss). Determine and record the percentage of efficiency using the values of R_E and $a^2 R_L$. Is this value larger than that recorded in Fig. 20-10? Would you expect it to be?

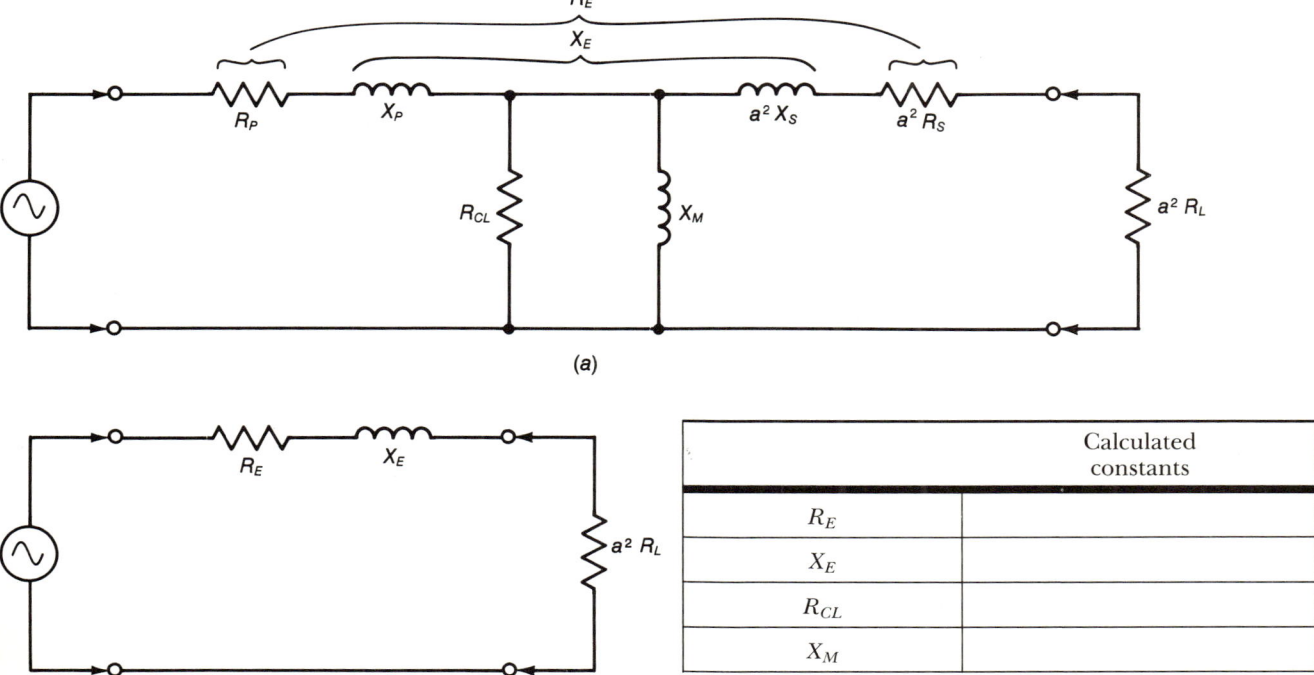

FIGURE 20-11   Equivalent circuit constants.

ACTIVITY 20-3   IMPEDANCE MATCHING

Introduction

Maximum transfer of power occurs when the resistances of the source and the load are matched. This activity will first verify that maximum transfer occurs when source and load resistances are equal. Then, the activity will show how a transformer can match unequal resistances or impedances.

Supplies

- (2) Transformers, dual 115-V primaries, dual 12-V 2-A secondaries (Triad F107Z or equivalent)
- (1) Fuse, 1-A, 3-AGC with holder
- (1) Resistor, 4.7-Ω, $\frac{1}{2}$-W, ±5%
- (1) Resistor, 10-Ω, $\frac{1}{2}$-W, ±5%
- (1) Resistor, 22-Ω, $\frac{1}{2}$-W, ±5%
- (1) Resistor, 27-Ω, $\frac{1}{2}$-W, ±5%
- (1) Resistor, 100-Ω, $\frac{1}{2}$-W, ±5%
- (1) Resistor, 220-Ω, $\frac{1}{2}$-W, ±5%
- (1) Resistor, 330-Ω, $\frac{1}{2}$-W, ±5%
- (1) Resistor, 1-kΩ, $\frac{1}{2}$-W, ±5%
- (2) Resistors, 2.2-kΩ, $\frac{1}{2}$-W, ±5%
- (2) Resistors, 5.1-kΩ, $\frac{1}{2}$-W, ±5%
- (1) Capacitor, 100-μF, *nonpolarized*, 10-V, ±10%
- (1) Inductor, 0.4-H, 22-Ω DCR, 275-mA (Stancor 6-2725 or quivalent)
- (1) DMM or VOM

Procedure

➥ **CAUTION:** The primary voltage used in this activity can cause a fatal shock. Use safe construction practices in connecting this circuit. Do not touch any part of the primary circuit after power is applied.

1. Transfer of power
   a. The transformer and resistor inside the dotted lines in Fig. 20-12 represent a source with 2.2 kΩ of internal resistance. This value of internal resistance was chosen so that low-power resistors could be used for this activity. (The internal resistance of the transformer itself is only a few ohms, about 24 Ω, as you found out in the previous activity. Thus, matching resistances would require the use of large power resistors.)
   b. Construct the circuit of Fig. 20-12 using the 1-kΩ resistor for R_L. Apply power and measure the voltage across R_L. Record this measured voltage in the V_{R_L} column of the table in Fig. 20-13. Now using the formula $P = V^2/R$, calculate, and record in the table, the value for P_{R_L} when R_L is 1 kΩ.
   c. Turn off the power. Change the load resistor to 2.2 kΩ and apply power. Make the measurement and calculation needed to complete the second row of the table in Fig. 20-13.
   d. Repeat step c above using a 5.1-kΩ load resistor.
   e. Which of the three loads dissipated the most power? Why?
   f. Do the data in Fig. 20-13 support the maximum power transfer theorem?

2. Resistance matching
   a. The circuit in Fig. 20-14 shows a transformer being used to match a low-resistance load to a higher resistance source. (Note that the source resistance is 220 Ω rather than the 2200 Ω used in the previous step.) With the terminal connections shown for the matching transformer, its voltage ratio is 115 V/24 V. Calculate and record its impedance ratio.
   b. Which of the load resistors listed in Fig. 20-15 should provide the best match for the source? Why?
   c. Construct the circuit shown in Fig. 20-14 and, using the procedures followed in step 1, complete the data table in Fig. 20-15.
   d. Do the data collected in step 2c support your answer to step 2b?
   e. If the load had been connected to only one of the secondaries in Fig. 20-14, would the 4.7-Ω or the 10-Ω load provide the better power transfer? Why?

3. Impedance matching
   a. Figure 20-16(a) shows a source that has an internal impedance rather than pure resistance. To match this internal impedance, the reflected load impedance must be the conjugate (same except for the sign of the angle) of the internal impedance. A few calculations, with some estimated values, will show this to be the case (approximately) when the load is a 27-Ω resistor paralleled by a 100-μF capacitor.

   Figure 20-16(b) is an equivalent circuit for Fig. 20-16(a). The values shown for the transformer constants are typical of those determined for this transformer by the open-circuit and short-circuit tests in the previous activity. In that activity, it was also determined that the transformation ratio (a) is 4.4 for this transformer.

   Using the transformation ratio and the value of R_L and C_L, the reflected impedance ($Z_L a^2$) can be determined as follows

   $$X_C = \frac{1}{2\pi f C} = \frac{1}{6.28 \times 60 \times 100 \times 10^{-6}} = 26.5 \; \Omega \; \underline{/-90°}$$

NAME _____ DATE _____

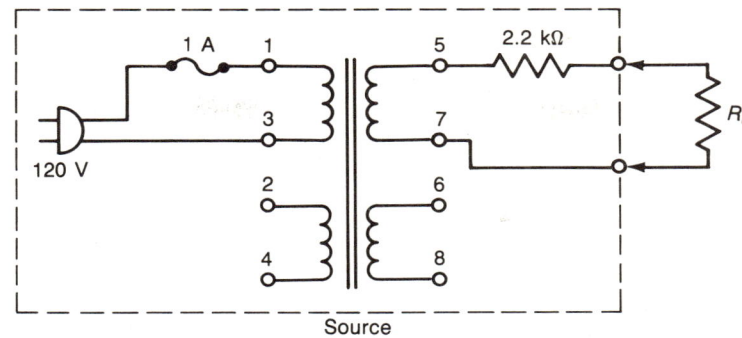

FIGURE 20-12 Source with 2.2 kΩ of internal resistance.

Load resistor R_L	Load voltage V_{R_L}	Load power P_{R_L}
1 kΩ		
2.2 kΩ		
5.1 kΩ		

FIGURE 20-13 Data table for step 1.

$$B_C = \frac{1}{X_C \angle -90°} = 0.038 \text{ S} \angle 90°$$

$$G = \frac{1}{R} = 0.037 \text{ S} \angle 0°$$

$$Y = B_C + G = 0.053 \text{ S} \angle 46°$$

$$Z_L = \frac{1}{Y} = 18.9 \text{ Ω} \angle -46°$$

$$Z_L a^2 = 4.4 \times 4.4 \times 18.9 \text{ Ω} \angle -46°$$
$$= 366 \text{ Ω} \angle -46°$$

Finally, the reactance of the inductor was found by

$$X_L = 2\pi f L = 6.28 \times 60 \times 0.4 = 151 \text{ Ω}$$

Compared to the other values in Fig. 20-16(b), R_E and X_E are very small. Therefore, they are omitted from the simpler equivalent circuit of Fig. 20-16(c).

Notice in Fig. 20-16(c) that the values of X_L, X_M, and R_{CL} are different than the values given in Fig. 20-16(b). The new values in Fig. 20-16(c) are more appropriate because both the transformer and the inductor are operating at only a small fraction of their rated currents. For example, the inductor is rated at 0.4 H when the current is fluctuating above and below 275 mA. In this circuit, the current through the inductor is less than 20 mA ac. At this reduced current, the inductor has about 0.8 H of inductance. Likewise, the reduced primary exciting current

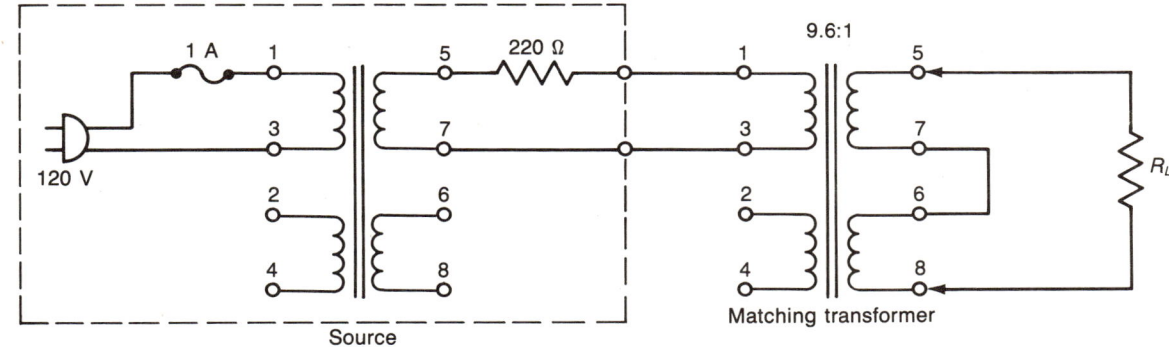

FIGURE 20-14 Reflected resistance.

Load resistor R_L	Load voltage V_{R_L}	Load power P_{R_L}
4.7 Ω		
10 Ω		
22 Ω		

FIGURE 20-15 Data table for step 2.

Copyright © 1993 by the Glencoe Division of Macmillan/McGraw-Hill School Publishing Company. All rights reserved.

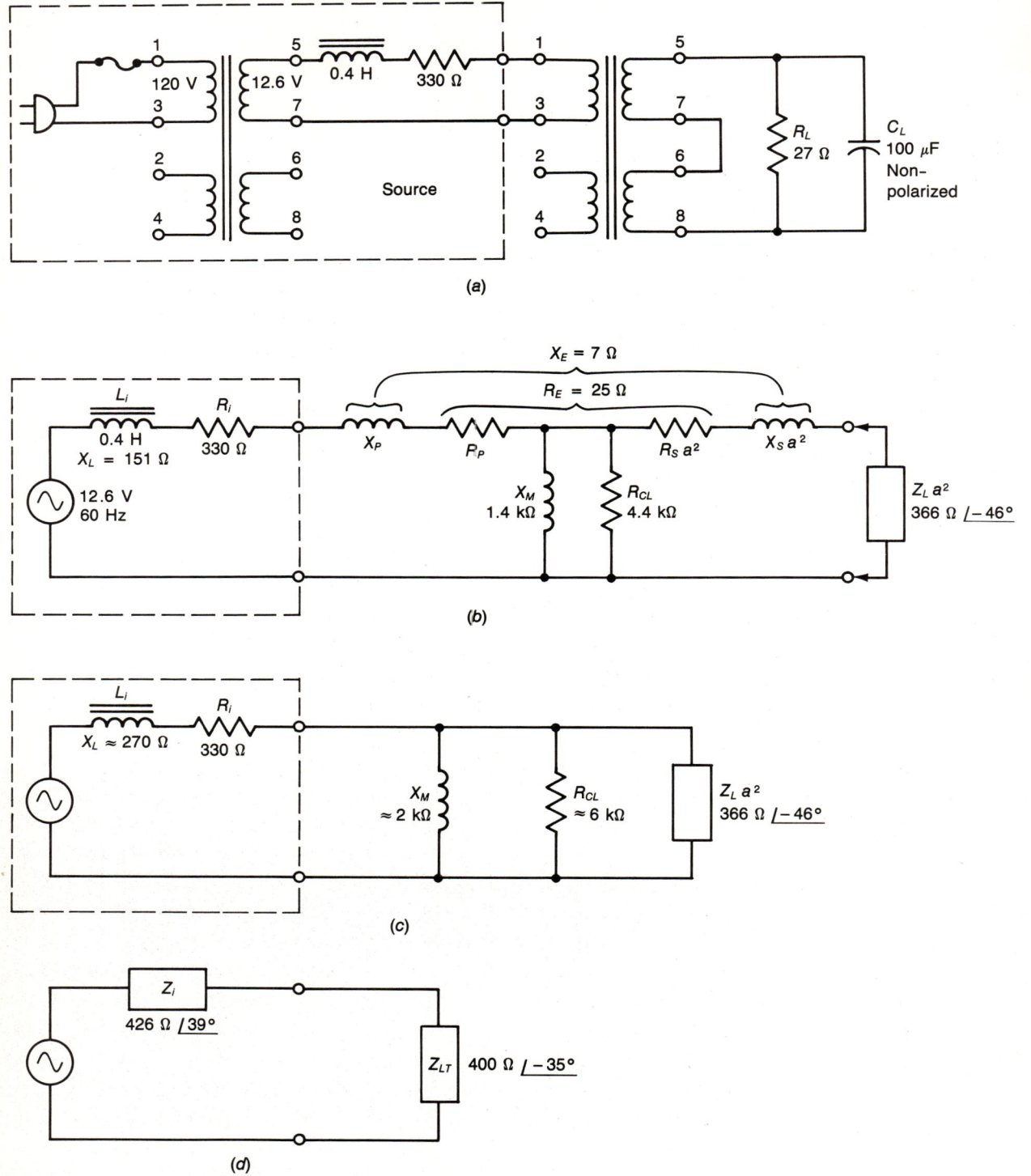

FIGURE 20-16 Circuits for Activity 20-3, step 3.

causes X_M to increase and R_{CL} to increase. (R_{CL} increases because the core losses decrease as the exciting current decreases.) The values of X_L, X_M and R_{CL} specified in Fig. 20-16(c) were determined experimentally by measuring the appropriate currents, voltages, and power.

Further study of Fig. 20-16(c) will show that the values of X_M and R_{CL} are small enough compared to $Z_L a^2$ that they must be included in the total load "seen" by the source. The total effective impedance (Z_{L_T}) can be determined as follows:

Load components		Resulting power	
R_L	C_L	V_{R_L}	P_{R_L}
10 Ω	0		
10 Ω	100μF		
27 Ω	0		
27 Ω	100μF		
100 Ω	0		
100 Ω	100μF		

FIGURE 20-17 Data table for step 3.

$$Y_{L_T} = \frac{1}{X_m} + \frac{1}{R_{CL}} + \frac{1}{Z_L a^2} = 0.5 \text{ mS } \underline{/-90°}$$
$$+ 0.2 \text{ mS } \underline{/0°} + 2.7 \text{ mS } \underline{/46°} = 2.5 \text{ mS } \underline{/35°}$$
$$Z_{L_T} = \frac{1}{Y_{L_T}} = 400 \text{ Ω } \underline{/-35°}$$

The internal impedance (Z_i) of the source can be estimated using the value of X_L given in Fig. 20-16(c). The calculations are

$$Z_i = \sqrt{X_L^2 + R_i^2} \underline{/\tan^{-1} X_L/R} = 425 \text{ Ω } \underline{/39°}$$

Using the values of Z_i and Z_{L_T} determined above, the final equivalent circuit can be drawn as in Fig. 20-16(d). This circuit shows that a fairly close impedance match occurs when the transformer in Fig. 20-16(a) is loaded by a 27-Ω resistor paralleled by a 100-μF capacitor.

b. Construct the circuit of Fig. 20-16(a). *Be certain that the 100-μF capacitor is a nonpolarized capacitor.* Apply power. Measure and record in Fig. 20-17 the voltage across R_L. Also calculate the load power by $P = V^2/R_L$. Will removal of C_L increase or decrease the power delivered to the load? Why? Check your answer by removing C_L and completing the third row of Fig. 20-17.

c. Make the necessary circuit changes, voltage measurements, and calculations needed to complete Fig. 20-17. Which load provides the best impedance match?

d. Is the phase angle of the 100-Ω, 100-μF load combination larger than or smaller than the phase angle of the source impedance (Z_i)? Using the procedure outlined in step 3a, calculate and record Z_{L_T} for this load combination. Was your answer to the first question correct?

ACTIVITY 20-4 SERIES-AIDING AND SERIES-OPPOSING WINDINGS

Introduction

A multiple-secondary, multiple-primary transformer can provide many voltages besides those specified for the individual windings. A transformer's windings can be connected in any configuration that does not exceed the volts-per-turn ratio or the current rating of any winding.

Supplies

(1) Transformer, dual 115-V primaries, dual 12-V, 2-A secondaries (Triad F107Z or equivalent)
(1) DMM or VOM
(1) Fuse, 1-A, 3-AGC with holder

Procedure

◆ **Caution:** The voltages used in this activity can cause a fatal shock. Use safe construction practices in connecting your circuits. Do not touch any part of the circuit after power is applied.

1. In this activity, the transformer will not be loaded and will be intermittently powered; thus, it is permissible to use the 120 V available from a standard outlet to power a 115-V winding.

2. Assume that the outlet voltage in your laboratory is 120 V. Ascertain the winding connections and phasings needed to provide as many output voltages as possible. You should be able to produce 14 output voltages ranging from 6 V to 145 V. Remember that you cannot connect one of the 12-V windings series-opposing one of the 115-V windings and apply 120 V to the series combination. This would exceed the volts-per-turn rating of the windings.

3. Construct and test the transformer configurations you have designed.

◆ **Caution:** Remember to remove power from the circuit before changing connections to the windings.

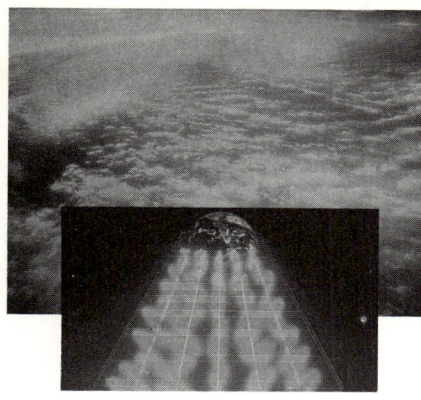

Chapter 21

Measuring Instruments

ACTIVITY 21-1 METER TOLERANCE AND LOADING

Introduction

The purpose of this activity is to investigate the loading effects of the DMM and the VOM. The data collected will show that the input resistance of a VOM changes with the range setting while the input resistance of a DMM is independent of the range setting.

Supplies

(1) Power supply, 0- to 25-V dc
(2) Resistors, 10-kΩ, $\frac{1}{4}$-W, ±5%
(2) Resistors, 910-kΩ, $\frac{1}{4}$-W, ±5%
(1) VOM, 20-kΩ/V sensitivity on dc voltage ranges
(1) DMM, 10-MΩ input resistance
(1) Signal generator, audio range

Procedure

1. Dc voltmeter loading
   a. Construct the circuit shown in Fig. 21-1(a). Set the voltage from the power supply to 3.00 V as indicated by the DMM.
   b. Calculate, and record in Fig. 21-1(b), the voltages across R_1 and R_2. Use the nominal values given on the schematic diagram.
   c. Make and record the measurements needed to complete the data table in Fig. 21-1(b). Read the VOM scale to the nearest division.
   d. Calculate and record the minimum voltage across R_1 or R_2 under worst-case tolerance conditions; i.e., one resistor at +5 percent, the other resistor at −5 percent.
   e. Assume the VOM is reading on the low end of its tolerance (−2 percent). If the meter did not load the circuit, what would be the minimum measured voltage across R_1 or R_2 under worst-case conditions of both meter and resistance tolerance?
   f. Considering the tolerances of the resistors and meter, is there any evidence of significant meter

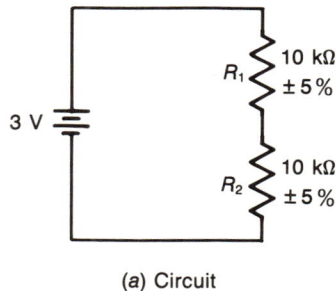

(a) Circuit

	Calculated voltage	Measured voltage			
		VOM		DMM	
		2.5-V range	10-V range	2-V range	20-V range
V_{R_1}					
V_{R_2}					
V_T	3 V	✕		✕	3 V

(b) Data table

FIGURE 21-1 Meter loading in a dc circuit.

loading from the data you collected for Fig. 21-1? If yes, when? Explain.

g. Are the voltages you measured on the 10-V range of VOM within the expected range of voltage considering resistor and meter tolerances?

h. If the resistors in Fig. 21-1(a) were 910 kΩ, which meters and ranges would you expect to cause significant loading?

i. Change the resistors to 910 kΩ and then check your answer to step 1h. How much voltage did you measure across R_1 with the VOM on the 2.5-V range?

2. AC voltmeter loading

a. Construct the circuit shown in Fig. 21-2(a). Set the generator frequency controls to 20 Hz. Set the generator output to 3 V as measured by DMM on the 3-V range.

b. Using the meters and ranges indicated, complete row 1 of the table in Fig. 21-2(b). Which meter indicates the greater loading effect?

c. Change the generator's output frequency to 200 Hz. The output voltage of a generator often changes when the frequency is changed. Therefore, use the DMM to check the generator's output voltage and adjust for 3 V if necessary.

d. Now complete the second row of the table in Fig. 21-2(b).

e. Complete the rest of the data table in Fig. 21-2(b). Be sure to check the generator's output voltage, and adjust if necessary, every time the frequency is changed.

f. Does the loading effect become more severe as the frequency increases?

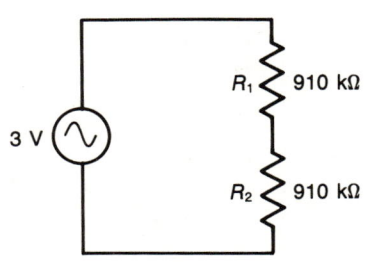

(a) Circuit

f	V_{R_1} measured	
	VOM 2.5-V range	DMM 2-V range
20 Hz		
200 Hz		
2 kHz		
20 kHz		

(b) Data table

FIGURE 21-2 Meter loading in an ac circuit.

ACTIVITY 21-2 SHUNTS, MULTIPLIERS, AND OHMMETERS

Introduction

In this activity you will convert a panel (single-range) microampere meter into a 10-V voltmeter, a 10-mA ammeter, and an ohmmeter. After constructing the ohmmeter circuit, you will check its scale calibration at several points.

Supplies

(1) Power supply, 0- to 25-V dc
(1) Panel meter, 100-μA, 650-Ω internal resistance (diode protection recommended)
(1) Resistor, 6.49-Ω, $\frac{1}{2}$-W, ±1%
(1) Resistor, 71.5-Ω, $\frac{1}{2}$-W, ±1%
(1) Resistor, 100-kΩ, $\frac{1}{2}$-W, ±1%
(1) Resistor, 1-kΩ, $\frac{1}{2}$-W, ±5%
(1) Resistor, 1.5-kΩ, $\frac{1}{2}$-W, ±5%
(1) Resistor, 3-kΩ, $\frac{1}{2}$-W, ±5%
(1) Potentiometer, 1-kΩ, $\frac{1}{2}$-W, ±5%
(1) DMM or VOM

Procedure

1. Voltmeter/multiplier

a. Calculate and record the value of the multiplier needed to convert the panel meter specified in the supplies to a 10-V voltmeter.

b. From the resistors listed under supplies, select the nearest value to the value calculated in step 1a. Use this value and the panel meter to construct a 10-V voltmeter. Now, test the accuracy of this meter by comparing its readings with those of the DMM. Do this by connecting both meters to the variable power supply and adjusting the supply output for 9 V as indicated on the DMM. How much voltage is indicated by the meter you constructed when the DMM indicates 9 V? Check your meter against the DMM at 1, 4, and 7 V.

c. Assuming the DMM is accurate, what was the largest percentage of error in the four readings you checked in step 1b? At what voltage was the error greatest?

2. Ammeter/shunt

a. Calculate and record the value of shunt needed to convert the 100-μA panel meter to a 10-mA ammeter.

b. Construct the 10-mA ammeter and connect it in the circuit shown in Fig. 21-3(a). Adjust the power supply for the DMM currents shown in

NAME _____   DATE _____

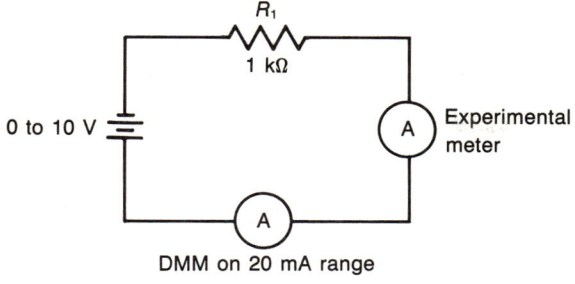

(a) Test circuit

DMM reading	Exp. meter reading
1.0 mA	
2.0 mA	
3.0 mA	
4.0 mA	
5.0 mA	
6.0 mA	
7.0 mA	
8.0 mA	
9.0 mA	
10.0 mA	

(b) Data table

FIGURE 21-3 Checking the 10-mA ammeter against the DMM.

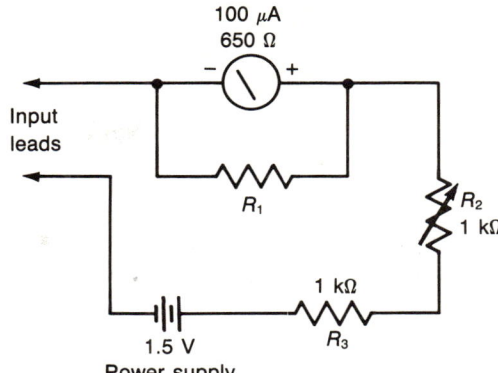

FIGURE 21-4 Experimental ohmmeter circuit.

Fig. 21-3(b) and record the corresponding readings for the meter (experimental) you constructed.

c. From the data you collected in step 2b, determine the percentage of error (worst case) for your meter, assuming the DMM is accurate.

d. Compared to the tolerance of the resistor in the test circuit, is your meter causing any significant loading effect? Check your answer by shorting out your meter with a jumper wire. Did the DMM reading change by more than 1 percent?

3. Ohmmeter circuit

   a. In the ohmmeter circuit shown in Fig. 21-4, the power supply is serving as a 1.5-V cell. Adjust its output for 1.50 V, as indicated by the DMM, and leave it at this setting for the remainder of this activity.

   This ohmmeter circuit is to have a center-scale reading of 1.5 kΩ. How much total internal resistance will the ohmmeter circuit have? How much current will flow when the ohmmeter is ohms-adjusted for zero ohms?

   b. Calculate and record the value of R_1 needed to shunt the 100-μA meter movement so that the circuit can handle the current determined in step 3a.

   c. Select from the supplies list the resistor with a resistance nearest to the value you calculated in step 3b and construct the ohmmeter circuit shown in Fig. 21-4. How much current should the meter movement indicate when measuring zero ohms with this circuit? Short the input leads of the circuit together and adjust R_2 to obtain this amount of current. The meter is now ohms-adjusted.

   d. How much current should the meter movement indicate when the ohmmeter circuit is measuring 1500 ohms? Insert a 1.5-kΩ resistor between the input leads and record the current indicated on the meter movement. Determine and record the percentage of error between the expected and the measured currents.

   e. How much resistance should be represented by one-quarter scale current with this ohmmeter circuit? Insert this value of resistance between the input leads and record the indicated current. Did the expected and measured current agree within 10 percent?

   f. Explain how you could modify this circuit so that center-scale resistance would be 150 Ω.

ACTIVITY 21-3   OSCILLOSCOPE MEASUREMENTS

Introduction

Through this activity you will gain experience in measuring phase and frequency with an oscilloscope. Also, you will demonstrate the advantage of using a ×10 probe for certain measurements.

Supplies

(1)   Signal generator, audio range
(1)   Oscilloscope with ×1 and ×10 probe
(1)   Capacitor, 0.1-μF, 25 DCWV
(1)   Resistor, 1-kΩ, ½-W, ±5%

(2) Resistors, 68-kΩ, ½-W, ±5%

(1) Transformer, dual 115-V primaries, dual 12-V, 2-A secondaries (Triad F107Z or equivalent)

Procedure

1. Phase measurement
   a. The circuit in Fig. 21-5 will provide voltages that are out-of-phase. The amount of phase shift between V_R and V_T can be measured with a single-trace oscilloscope if one remembers to use the external trigger. By using the external trigger connection shown in Fig. 21-5, the sweep will be triggered at the same point on the V_T waveform regardless of the waveform being displayed on the CRT screen.
   b. So that you will be able to check on the reasonableness of the phase shift you measure with the oscilloscope, first calculate and record the phase difference (shift) between V_T and V_R in the circuit in Fig. 21-5. Also draw an accurate phasor diagram for the circuit. The needed formulas are:

   $$X_C = \frac{1}{2\pi fC}$$

   $$\theta = \arctan\frac{X_C}{R}$$

   c. Construct the circuit shown in Fig. 21-5. With a ×1 probe, connect the vertical input of the oscilloscope to the top end of C_1. Adjust the oscilloscope controls to obtain a vertically centered display with one-half cycle spanning six major divisions on the X axis of the graticule. Mentally note the shape and location of the display so that you can compare it to the next display. How many degrees does one minor division on the X axis represent?
   d. When you move the vertical input to the top end of R_1, will the displayed waveform be shifted to the left or to the right of the present display? How many divisions? Check your answer by moving the vertical input to the top end of R_1.
   e. Measure and record the amplitude (peak-to-peak) and the period of the voltage across R_1.

2. Measuring frequency
   a. The circuit in Fig. 21-6 will display a Lissajous pattern on the screen of the CRT. The shape of the pattern will depend on the phase shift between V_T and V_R. Calculate and record the phase angle between V_T and V_R. Describe the shape of the Lissajous pattern this circuit should provide. How long should it take the electron beam in the CRT to trace this pattern one time?
   b. Construct the circuit in Fig. 21-6 and adjust the oscilloscope controls to obtain the expected pattern. (Remember to put the calibrated sweep control on the H [or X] input position.) Now connect the output of the signal generator to the Z axis input (intensity modulation input) of the oscilloscope. (The ground of the signal generator is connected to the ground of the oscilloscope.) Set the generator's frequency for 180 Hz and adjust its output level until the display shows signs of breaking into segments. Then carefully adjust the generator's frequency to obtain a stationary pattern. How many segments is the display chopped into?
   c. Switch the generator's function switch back and forth between sine wave and square wave. Describe the difference in the observed waveform when the generator's output is a square wave rather than a sine wave. What causes this difference?
   d. At what frequency should the display be chopped into 10 segments?
   e. Check the calibration of the generator's frequency dial at the highest frequency that you can.

3. The ×10 probe
   a. Construct the circuit shown in Fig. 21-7(a). Set the generator's frequency for 2 kHz and adjust its output controls for 10 V_{p-p} as indicated by the oscilloscope with the ×10 probe. Now make the measurements necessary to complete the top row of the table in Fig. 21-7(b). Is there any evidence that either probe is loading the circuit? If so, what is the evidence?
   b. Change the generator's frequency to 20 kHz and adjust, if necessary, the output voltage for 10 V_{p-p}. Then complete the second row of the data table. What evidence, if any, is there that the circuit is being loaded by either, or both, probes?
   c. Make the necessary changes and measurements needed to complete the last row in the data table. What evidence of circuit loading is there at this frequency?

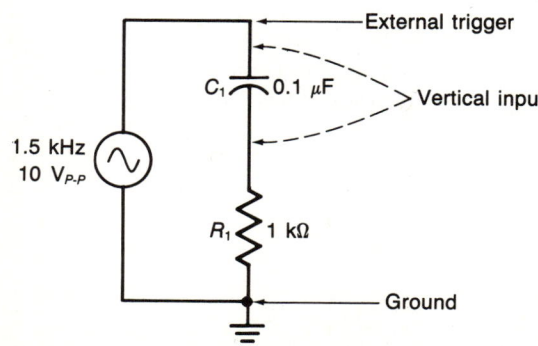

FIGURE 21-5 Circuit for measuring phase shift.

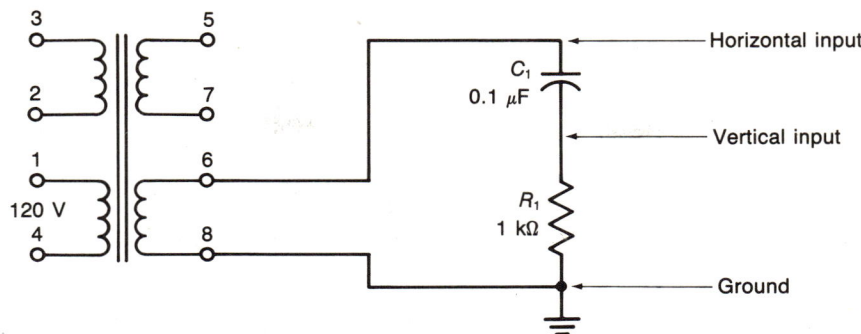

FIGURE 21-6 Displaying a Lissajous pattern.

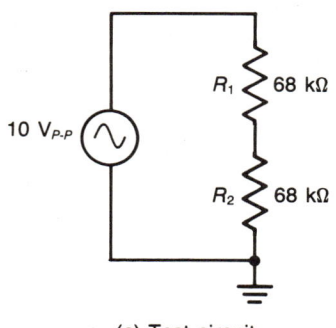

(a) Test circuit

f	V_T (p-p)		V_{R_2} (p-p)	
	×10 probe	×1 probe	×10 probe	×1 probe
2 kHz	10.0 V			
20 kHz	10.0 V			
100 kHz	10.0 V			

(b) Data table

FIGURE 21-7 ×10 probe versus ×1 probe.

ACTIVITY 21-4 DESIGNING AN ANALOG METER

Introduction

A single-range panel meter can be used to measure voltage, current, or resistance by adding appropriate amounts of resistance and voltage in the meter circuit. The meter circuit you design in this activity will have limited accuracy because you will have to use combinations of nonprecision resistors in your meter circuit.

Supplies

(1) DMM
(1) Power supply, 0- to 25-V dc
(1) Panel meter, 100-μA, 650-Ω internal resistance (diode protection recommended)

Miscellaneous resistors selected from the Materials List in this manual.

Procedure

1. Design, construct, and test meter circuits that convert the panel meter to:
   a. A 25-V voltmeter
   b. A 5-mA ammeter
   c. A 600-Ω center-scale ohmmeter that can be powered with one or more D cells.
2. Following the guidelines set forth in Appendix 1, prepare a report about this activity.

NAME _____ DATE _____

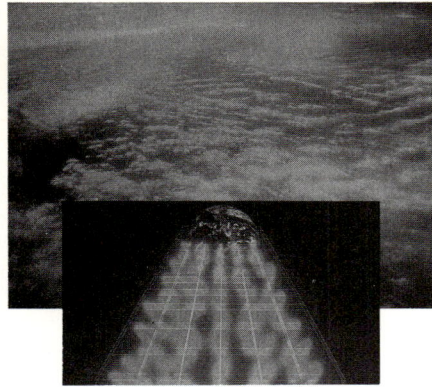

Chapter 22

Three-Phase Circuits

ACTIVITY 22-1 THREE-PHASE CONNECTIONS

Introduction

In this activity you will connect transformer primaries and secondaries in delta and wye configurations. You will verify the current and voltage relationships for the delta and wye connections when the load is balanced. You will determine the neutral-wire current under various load conditions. Finally, you will determine load power using several different approaches.

Supplies

- (3) Transformers, dual 115-V primaries, dual 12-V, 2-A secondaries (Triad F107Z or equivalent)
- (3) Lamps, 14-V, 0.08-A (no. 756 or equivalent) with holders
- (3) Fuses, 1-A, 3-AGC with holders
- (1) Multimeter (DMM) with 15-V, 100-mA and 200-mA ac ranges
- (1) Resistor, 220-Ω, 1-W
- (1) Oscilloscope with external triggering and ×1 probe.

Procedure

1. Delta-delta
   a. Figure 22-1(a) shows a three-phase circuit which uses three single-phase transformers to step 208-V, 3-ϕ down to 12-V, 3-ϕ. In this figure, each transformer has a dual 115-V primary winding. This dual winding is connected series-aiding so that the effective primary voltage is 230 V. Thus, this bank of three single-phase transformers can be connected to any three-phase outlet that provides between 208 and 230 V. Of course, the secondary voltages will be about 10 percent higher when the primaries receive 230 V than when they receive 208 V.

   It is extremely important that the primaries be wired exactly as shown in Fig. 22-1(a). If either of the dual primaries on one of the transformers is reversed, the flux of one primary will counter the flux of the other primary. This, of course, short-circuits two of the 208-V lines and causes one or more fuses to blow. If either the primaries or the secondaries are incorrectly phased, the delta will be short-circuited and the currents in the transformer windings will soar. When the incorrect phasing is in the primary, the fuses blow quite rapidly. However, when the incorrect phasing is in the secondary, the fuses may respond slowly, if at all, because the internal resistance of the transformer and the current step-down ratio of the transformer may not cause a very significant increase in the primary current.

   The primary connection shown in Fig. 22-1(a) will be used throughout the rest of this activity. The figures for the rest of the circuits will show only the secondary connections. The transformers used in this activity also have dual secondaries, but since only one secondary is used, only one secondary will be shown in the schematic diagrams.

   ⇨ **CAUTION:** The primary voltage used in this activity can cause a fatal shock. Make all connections to the circuit before plugging the circuit into the 3-ϕ 208-V outlet. Do not touch any part of the circuit while it is connected to the outlet.

   b. Construct the circuit shown in Fig. 22-1(a), but connect a voltmeter in the secondary delta as shown in Fig. 22-1(b). Apply power to the circuit. Read and record the voltage indicated by the voltmeter. Are the secondaries properly phased? Remove power from the circuit.

Copyright © 1993 by the Glencoe Division of Macmillan/McGraw-Hill School Publishing Company. All rights reserved.

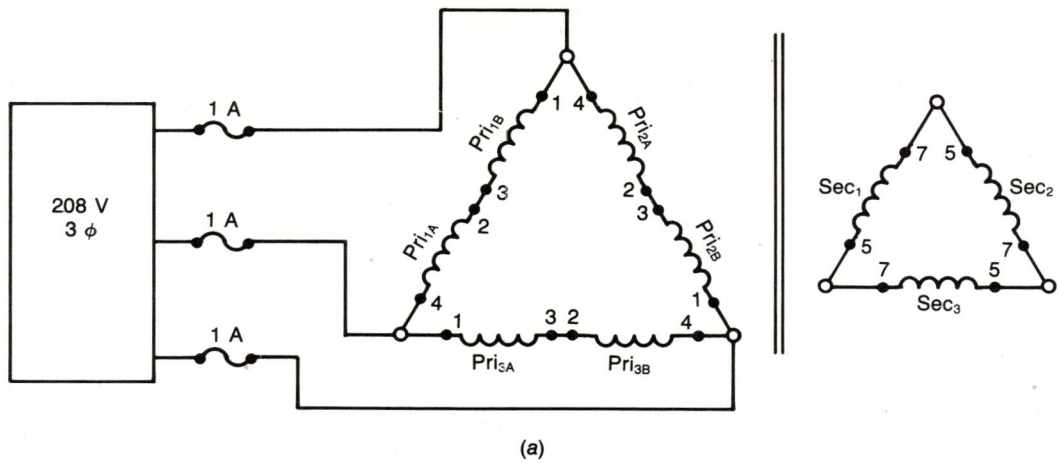

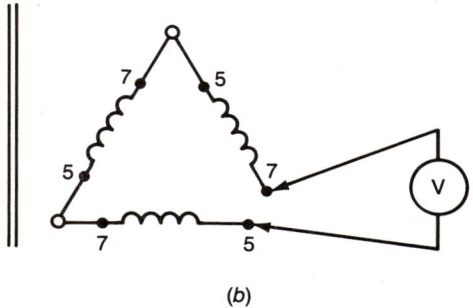
(b)

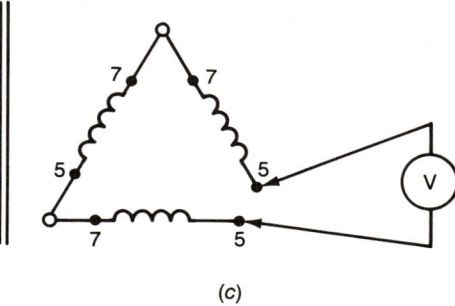
(c)

FIGURE 22-1 Phasing a delta secondary.

c. Reverse the secondary winding of transformer number 2 (Sec$_2$) as shown in Fig. 22-1(c). Apply power to the circuit. Read and record the indicated voltage. Are the secondaries properly phased? Remove power from the circuit.

d. Connect a three-lamp delta load to the delta secondary windings as shown in Fig. 22-2(a). Apply power to the circuit. Do the lamps glow with equal intensity? Note that all lamps may be dimmer than expected because the lamps are rated at 14 V and the transformers provide approximately 12 V at the secondary. Measure and record the voltage across each lamp. Remove the power from the circuit.

e. Compare the voltage measured in step 1c with the voltages measured in step 1d. Should they be equal? Why?

f. Figures 22-2(b) and (c) show the appropriate connections for measuring a load current and a line current, respectively. Remembering to remove power before making any circuit changes, measure (and record in the table in Fig. 22-3) the line currents and the load currents for the circuit in Fig. 22-2(a). Considering the tolerances of transformers and light bulbs, is this a balanced delta load?

g. Do the data collected in step 1f above indicate that the line current is 1.73 times as great as the load current for a balanced delta load?

2. Wye-wye
  a. Figure 22-4 shows a wye-connected source (transformer secondaries) connected to a balanced wye load. How much current should flow in the neutral wire if the lamps and the transformer secondaries are matched?
  b. Construct the circuit of Fig. 22-4 and then apply power. Do the lamps glow with equal brilliance? Remove power and then remove the neutral wire. Apply power again. Do the lamps still glow with equal brilliance? Remove power.
  c. Connect an ammeter between the star points (neutral points) of the source and the load. Apply power and read and record the current indicated by the ammeter. Does this measured current support your answer to step 2a? (Remember that the lamps and transformers are not perfectly balanced as indicated by the measured values in Fig. 22-3.) Remove power from the circuit but leave the ammeter connected to the star points.
  d. Connect a 220-Ω, 1-W resistor in parallel with L_1. How much current should this resistor draw when power is applied? How much current would you expect the ammeter to indicate when power is applied? Apply power. Read and record the current indicated by the ammeter. Remove power.

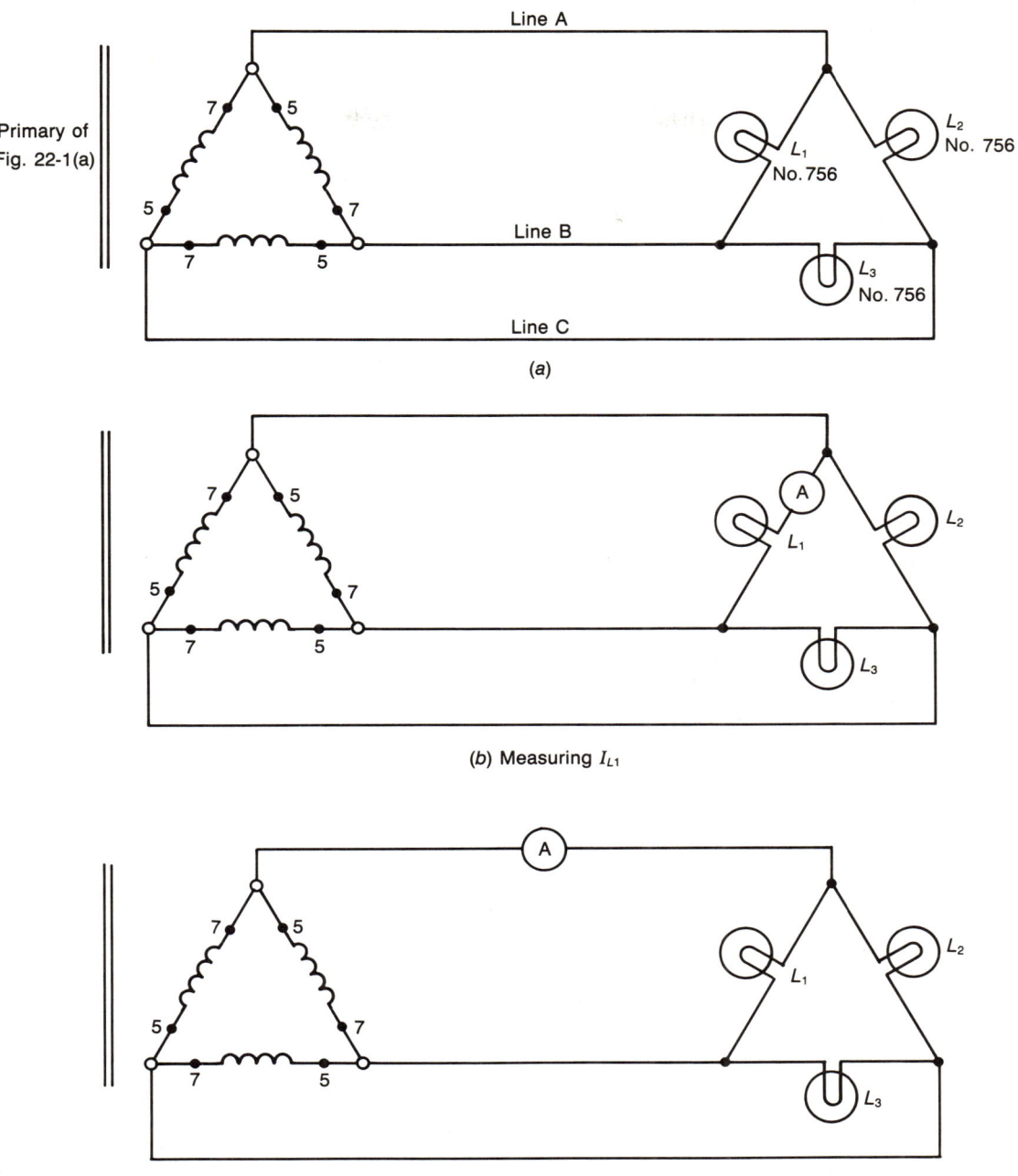

FIGURE 22-2 Balanced delta-load currents.

e. Remove the ammeter, but do not replace the neutral wire or remove the 220-Ω resistor. This leaves an unbalanced wye load without a fourth wire (neutral wire). When power is applied, should the lamps glow with equal intensity? Apply power. Now measure and record the voltage across each lamp. Remove power. Remove the 220-Ω resistor and connect the neutral wire between the star points.

f. The circuit should now be wired as shown in Fig. 22-4. Apply power and make the voltage measurements specified in Fig. 22-5. Record the measured voltages. Remove power. What relationship do the line voltages have to the phase voltages?

g. Now you will check the phase relationships between the three phase voltages. Connect the ground lead of the oscilloscope to the neutral point of the circuit. Connect a lead from the external trigger input jack to line A. Connect the vertical input probe to line A also. Apply power to the circuit and adjust the oscilloscope controls for a stable display of one cycle centered on the graticule. Sketch this waveform on the graph in

Measured	Currents
$I_{L_1} =$	
$I_{L_2} =$	
$I_{L_3} =$	
$I_{\text{line } A} =$	
$I_{\text{line } B} =$	
$I_{\text{line } C} =$	

FIGURE 22-3 Data table for Activity 22-1, step 1f.

Fig. 22-6. Do not adjust any of the oscilloscope controls for the following measurements. Move the vertical input probe to line B and observe the location of the waveform. How many degrees is this waveform displaced from the previous waveform? Also draw this waveform on the graph in Fig. 22-6. Now move the vertical input to line C and again record the waveform in Fig. 22-6. How many degrees is this waveform shifted from the last waveform? What is the phase relationship between the three phase voltages you measured? Remove power and disconnect the oscilloscope from the circuit.

3. Power relationships
   a. You will now use the data you have collected to solve for the power in a balanced delta load. The total power of a three-phase load is equal to the sum of the power of the individual phase loads. Since the lamps used in the circuit of Fig. 22-2(a) are resistive loads, you can use the voltages measured in step 1d and the currents measured in step 1f (and recorded in Fig. 22-3) to calculate the phase power. Calculate the three phase powers and record their values in Fig. 22-7. Add these three phase powers to obtain the total power and record the total power in Fig. 22-7.

   b. The total power of a three-phase load can also be determined by the line currents and line voltages. The two-wattmeter method of measuring power provides a power reading for each meter which is based on a line current and a line voltage. With a delta load, a line voltage is also a phase voltage. With a balanced resistive load the line current will be 30° out-of-phase with phase current. Since the phase voltage and phase current are in-phase (for a resistive load like lamps), the line current and phase voltage will be 30° out-of-phase. Therefore, the power indicated by a wattmeter connected to respond to line voltage and line current can be calculated by $P = I_{\text{line}} V_{\text{line}} \cos 30°$ when the load is resistive. Using this formula and the average line currents and line voltages from steps 1d and f, calculate the power a wattmeter would indicate. Record this value. According to the two-wattmeter method, doubling this power should yield the total power for a balanced load. Double this power and record the total power in the second column of Fig. 22-7.

   c. How does the estimate of total power obtained in step 3b above compare with that obtained in step 3a above?

Measured	Voltages
$V_{L_1} =$	
$V_{L_2} =$	
$V_{L_3} =$	
$V_{AB} =$	
$V_{BC} =$	
$V_{CA} =$	

FIGURE 22-5 Data table for Activity 22-1, step 2f.

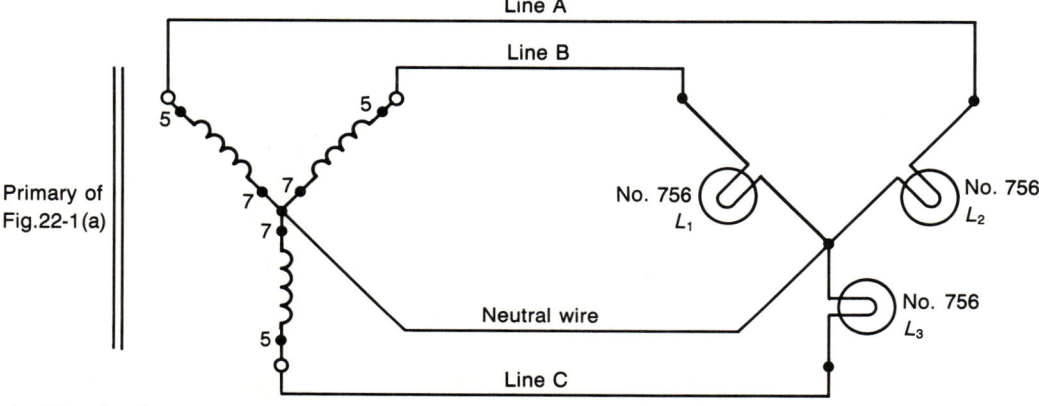

FIGURE 22-4 Wye load.

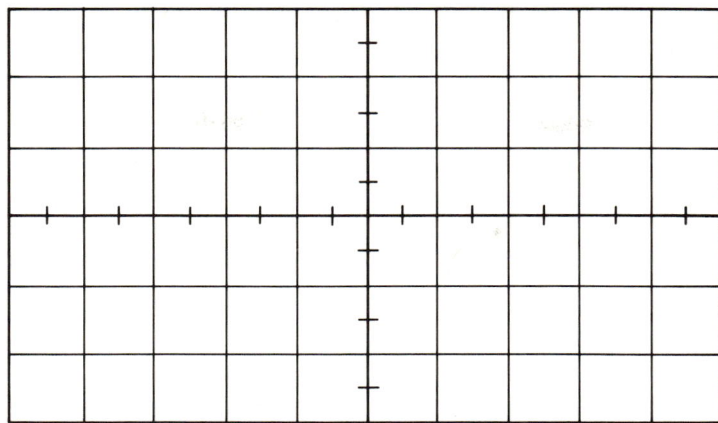

FIGURE 22-6 Graph for oscilloscope waveforms.

	P_T using I_P and V_P	P_T using I_L and V_L
P_{L_1}		
P_{L_2}		
P_{L_3}		
P_T		

FIGURE 22-7 Data table for Activity 22-1, step 3a.

ACTIVITY 22-2 UNBALANCED WYE LOADS

Introduction

In a four-wire wye system, the amount of neutral-wire current is determined by the magnitude and phase of the wye loads. Any time the loads are not balanced, in either magnitude or phase, some current flows in the neutral wire. In this activity you will need to determine the components required to produce the specified loads, calculate the currents, and construct and test the circuit you have designed. Remember that the phase of a current can be measured by inserting a small resistance (small compared to other circuit impedances) in the appropriate line.

Supplies

(3) Transformers, dual 115-V primaries, dual 12-V, 2-A secondaries (Triad F107Z or equivalent)

(3) Fuses, 1-A, 3-AGC with holders

(1) DMM with ac ranges

(1) Oscilloscope with external triggering and ×1 probes

Miscellaneous resistors, capacitors, and inductors selected from the Materials List in this manual.

Procedure

↪ **Caution:** The voltages used in this activity can cause a fatal shock. Use safe construction practices in connecting your circuits. Do not touch any part of the circuit after power is applied.

1. Design, construct, and test a four-wire wye system with a line voltage of 20.76 V [see Fig. 22-1(a) for primary terminal connections and Fig. 22-4 for secondary terminal connections]. Loads for the system are to be 220 Ω $/0°$, 834 Ω $/10.4°$, and 1306 Ω $/-29.5°$.

2. Write a detailed activity report. Your report must meet or exceed the requirements outlined in Appendix 1.

Chapter 23

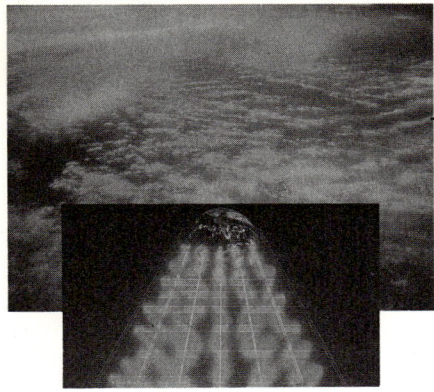

Introduction to Discrete Electronics

ACTIVITY 23-1   DIODE CHARACTERISTICS

Introduction

This activity will acquaint you with the characteristics of a silicon rectifier diode. With the data collected you will be able to produce an I-V curve for a diode and calculate the internal resistance (r_D) of a diode.

Supplies

(1) Power supply, 0 to 25-V dc, 0.5-A capacity, floating output
(1) Transformer, dual 115-V primaries, dual 12-V, 2-A secondaries (Triad F107Z or equivalent)
(1) Diode, IN4002 or equivalent
(1) Resistor, 1-Ω, ½-W, ±5%
(1) Resistor, 100-Ω, 20-W, ±5%
(1) Oscilloscope, 10-mV/cm vertical sensitivity
(1) Ammeter, dc, multirange (VOM or DMM)
(1) Voltmeter, dc, multirange, 10-MΩ input resistance (DMM or electronic MM)

Procedure

1. Look at Fig. 23-1. Is the diode forward-biased or reverse-biased? When the source voltage is 20 V will the current in the circuit be less than or greater than 1 mA?
2. Reverse-biased diode
   a. Construct the circuit in Fig. 23-1 and collect the data needed to complete the table in Fig. 23-2. *Note:* Remove the voltmeter from the circuit before reading the ammeter; otherwise, the ammeter will indicate I_D plus the current required by the voltmeter.

⚡ **Caution:** The 20-W resistor will get hot; do not burn your fingers!

   b. When a diode is reverse-biased, does $V_D \approx V_S$?
   c. Is I_D directly proportional to V_D when a diode is reverse-biased?
   d. Does your completed data table support your answers to step 1 above?
3. Forward-biased diode
   a. Reverse the diode in the circuit of Fig. 23-1 and complete the data table in Fig. 23-3. Again, remove the voltmeter before reading the current if the current is less than 5 μA.

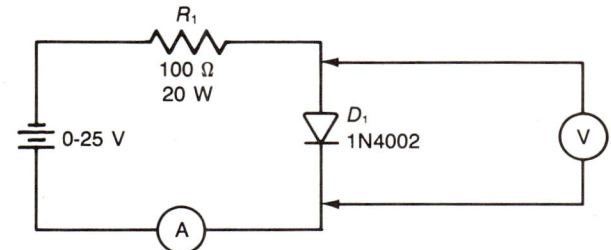

FIGURE 23-1   Circuit for determining diode characteristics.

V_S (volts)	V_D	I_D
0.5		
1.0		
2.0		
5.0		
10.0		
15.0		
20.0		
25.0		

FIGURE 23-2   Data table for the circuit in Fig. 23-1.

V_S (volts)	V_D	I_D
0.4		
0.5		
0.6		
0.7		
0.8		
0.9		
1.0		
2.0		
5.0		
10.0		
15.0		
20.0		
25.0		

FIGURE 23-3 Data table for circuit in Fig. 23-1 when the diode is reversed.

b. Do the data you collected indicate that a forward-biased diode is a linear device?

c. On a graph like that shown in Fig. 23-4, plot the I-V curve for a diode using the data you collected for Fig. 23-2 and Fig. 23-3.

d. Using your data and/or I-V curve, determine the diode's static resistance at $V_D = 0.55$ V.

e. Using your data and/or I-V curve, determine the diode's dynamic resistance as V_D changes from approximately 0.6 V to approximately 0.7 V.

4. Dynamic resistance of a diode
The circuit in Fig. 23-5 will be used to indirectly measure r_D (the dynamic resistance) of the diode. Remember that $r_D = \Delta V_D/\Delta I_D$. The ac voltage ($v_D$) measured across the diode is equal to ΔV_D. Also, $\Delta I_D = v_{R_1}/R_1$ where v_{R_1} is the ac voltage across R_1. Therefore,

$$r_D = \frac{v_D}{v_{R_1}/R_1} = \frac{v_D \times R_1}{v_{R_1}}$$

In Fig. 23-5, R_L limits the current to a safe value, and the 25-V dc source keeps the diode forward-biased at all times. The 12-V ac source causes the diode current and diode voltage to fluctuate around the value established by the 25-V dc source.

a. Construct the circuit in Fig. 23-5. *Turn on the dc supply before turning on the ac supply.* Measure and

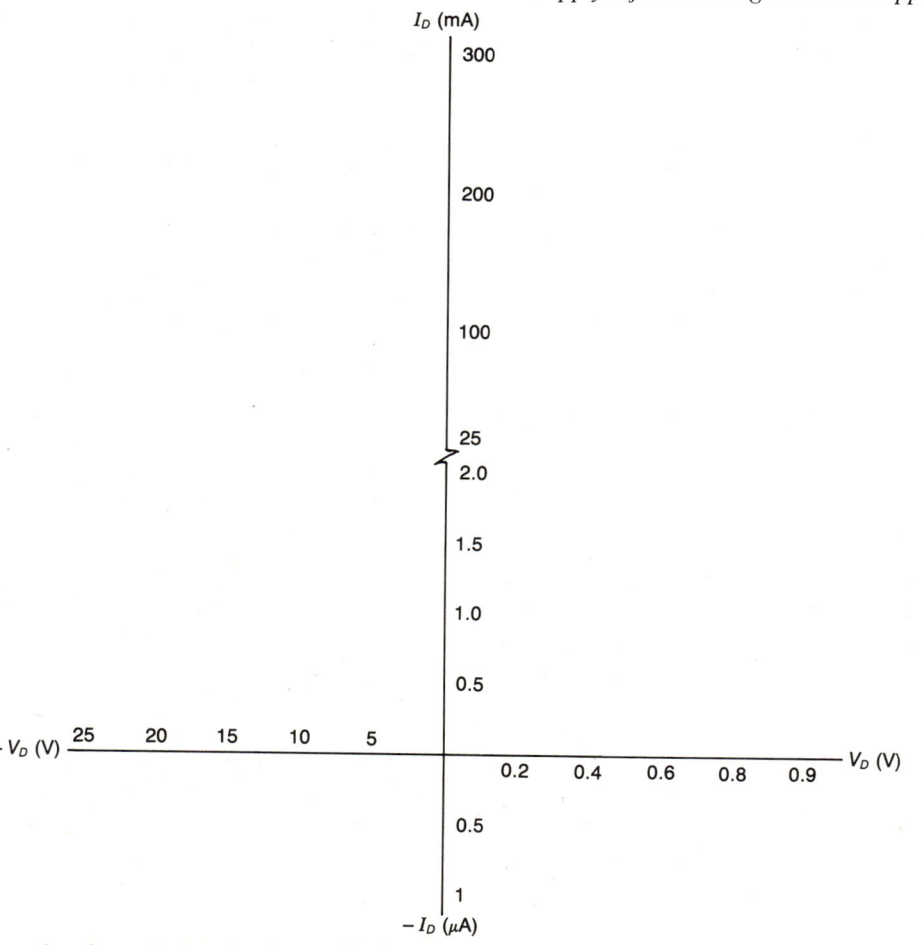

FIGURE 23-4 Example of graph for plotting a diode I-E curve.

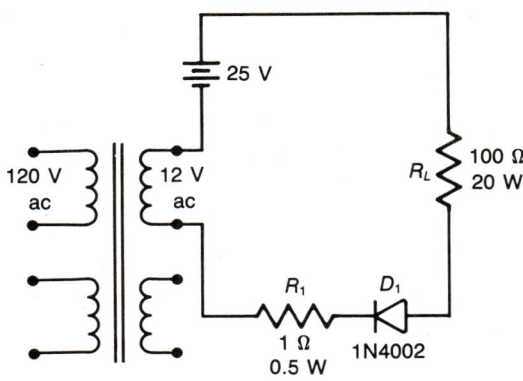

FIGURE 23-5 Circuit for determining r_D.

record V_{R_1} and V_D using the oscilloscope with ac coupling. Remove power from the circuit.
  b. Using the data just collected, calculate and record r_D.
  c. Does the value of r_D just calculated agree (within 20 percent) with the value of r_D calculated in step 3e? Should it? Why?
5. As a review, answer the following questions:
  a. To forward-bias a diode, should the anode be negative or positive with respect to the cathode?
  b. Does increasing the reverse voltage of a diode by a factor of 10 increase the leakage current by a factor of 10?
  c. What is the purpose of R_1 in Fig. 23-1?
  d. Will the static resistance (R_D) of a diode be greater at 1 mA or at 100 mA?
  e. Will the dynamic resistance (r_D) of a diode be greater when the current fluctuates around 1 mA or around 100 mA?

ACTIVITY 23-2 RECTIFIER CIRCUITS

Introduction

In this activity you will construct and test three common rectifier circuits: half-wave, full-wave with center-tapped transformer, and full-wave bridge. The data you collect in this activity will illustrate the major differences in these rectifier circuits.

Supplies

  (1) Transformer, dual 115-V primaries, dual 12-V, 2-A secondaries (Triad F107Z or equivalent)
  (4) Diodes, 1N4002 or equivalent
  (1) Resistor, 100-Ω, 20-W, $\pm 5\%$
  (1) Oscilloscope
  (1) Ammeter, dc, multirange (VOM or DMM)
  (1) Voltmeter, dc, multirange (VOM or DMM)

Procedure

1. Refer to Fig. 23-6.
  a. Which circuit (a, b, or c) will provide the least V_{dc}? Why?
  b. Which circuit (a, b, or c) will provide the most V_{dc}? Why?
2. Half-wave
  Construct the circuit shown in Fig. 23-6(a). Be certain the primary windings are connected properly.
  a. With the VOM or DMM, measure and record the values needed to complete column one in the table of Fig. 23-7. *Remember to put the ammeter in series* with R_L.
  b. With a calibrated dc-coupled oscilloscope, view and measure the output waveform across R_L. *Accurately* draw the observed waveform on a graph like that in Fig. 23-8(a).
  c. Next, view the waveform of the ac source (12-V secondary) and note its peak-to-peak value. How does this peak-to-peak value compare to the value of V_{ac} recorded in step 2a? How does it compare to the peak value of the waveform observed in step 2b?
  d. Calculate and record the values requested in the fourth column of the table in Fig. 23-7. For the most accurate results remember to take into account the voltage dropped across D_1.
  e. Compare the values measured in step 2a with the values calculated in step 2d. Are these values within the range of error that could be caused by the tolerance of your measuring equipment and the tolerance of R_L?
3. Full-wave center-tapped
  a. Study Fig. 23-6(b). Should the peak value of this circuit's output waveform be equal to, greater than, or less than that of Fig. 23-6(a)? Why?
  b. What relationship should exist between V_{dc} of Fig. 23-6(a) and V_{dc} of Fig. 23-6(b)?
  c. Construct the circuit of Fig. 23-6(b). Now make the measurements required to complete the second column of Fig. 23-7 and the waveform of Fig. 23-8(b).
  d. Do the measured values in step 3c support your answers to steps 3a and b?
  e. Calculate and record the values needed to complete column 5 in Fig. 23-7. Do your calculated values agree (within expected tolerances) with your measured values?
  f. What would be the effect on V_{dc} and the output waveform if terminals 5 and 7 on the transformer in Fig. 23-6(b) were interchanged? Why?
  g. Interchange terminals 5 and 7. Then check your answer to step 3f.
4. Full-wave bridge
  a. Should V_{dc} of Fig. 23-6(c) be greater than, less than, or equal to V_{dc} of Fig. 23-6(b)? Why?

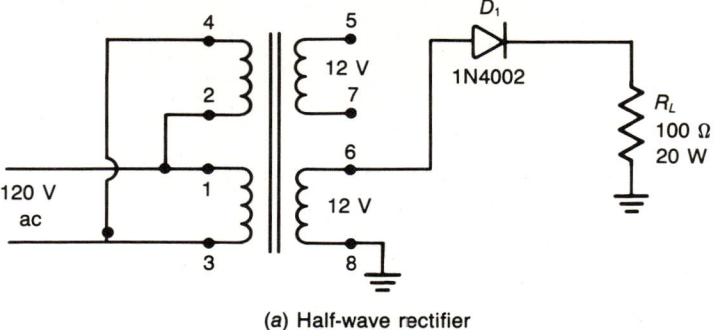

(a) Half-wave rectifier

(b) Full-wave center-tapped rectifier

(c) Full-wave bridge

FIGURE 23-6 Circuits for determining rectifier characteristics.

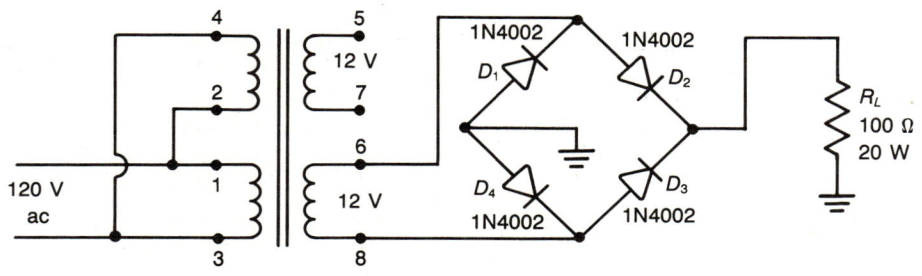

	Measured values (use VOM or DMM)			Calculated values (Use measured values of V_{ac})		
	Half-wave	Full-wave	Bridge	Half-wave	Full-wave	Bridge
V_{ac} (Sec)						
V_{dc} (V_{R_L})						
I_{dc} (I_{R_L})						

FIGURE 23-7 Data table for the circuits of Fig. 23-6.

b. Which diodes conduct at the same time in the circuit of Fig. 23-6(c)?

c. Construct the circuit shown in Fig. 23-6(c) and make the measurements needed to complete the third column in Fig. 23-7. Do your measured values recorded in Fig. 23-7 verify your answer to step 4a?

d. Make and record the calculations needed to fill in column 6 in Fig. 23-7. How closely do your calculated values agree with your measured values in step 4c?

5. As a review, answer the following questions:
  a. Assume D_1 in Fig. 23-6(a) is shorted. What value of V_{dc} would you expect? What would the output waveform look like?
  b. Assume D_1 in Fig. 23-6(b) is open. What value of V_{dc} would you expect? What would the output waveform look like?
  c. Assume both D_1 and D_2 in Fig. 23-6(c) are open. What value of V_{dc} would you expect?
  d. Suppose all three circuits in Fig. 23-6 had used both secondaries connected series-aiding as in

NAME _____ DATE _____

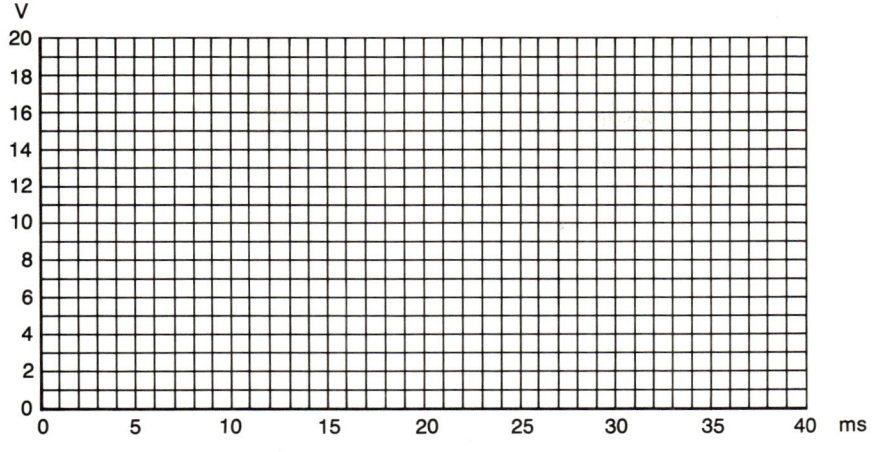

(a) Half-wave output waveform

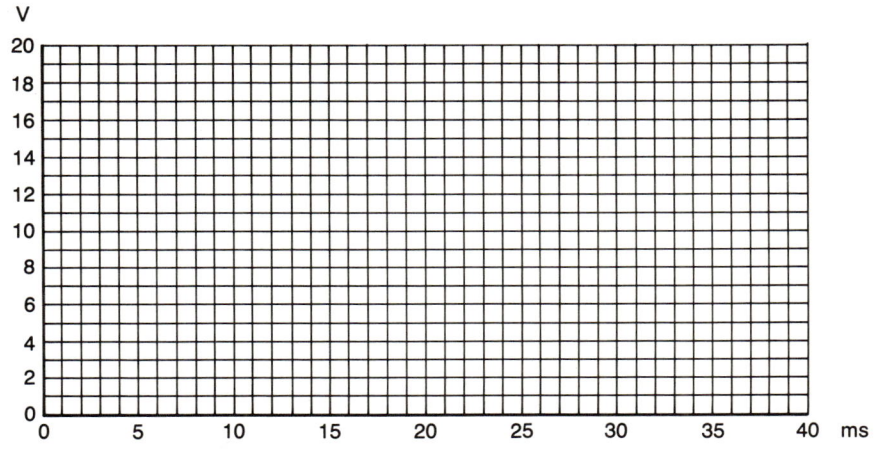

(b) Full-wave center-tapped output waveform

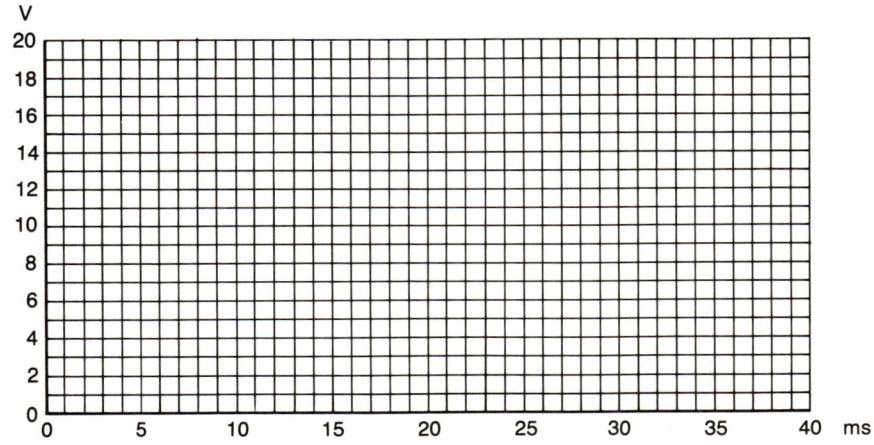

(c) Full-wave bridge output waveform

FIGURE 23-8 Graphs for plotting output voltage waveforms.

Copyright © 1993 by the Glencoe Division of Macmillan/McGraw-Hill School Publishing Company. All rights reserved.

113

Fig. 23-6(b). Which circuit would have provided the largest peak output voltage? Which circuit would have provided the most V_{dc}?

e. *Note:* Your answer to any of the above questions can be experimentally checked without harming any components or equipment.

ACTIVITY 23-3 TRANSFORMER INTERNAL RESISTANCE

Introduction

The internal resistance of a transformer is an important characteristic when the transformer is used to power a filtered power supply. It is one of the factors that determines the maximum surge current the diode must withstand. It causes the transformer's terminal voltage to decrease as the current increases. Thus, the peak voltage to which a filter capacitor can charge (and ultimately the value of V_{dc}) decreases as the peak repetitive current increases. This activity is concerned with determining the internal resistance of the transformer which will be used in the filtered power supplies in the next activity.

Supplies

(1) Transformer, dual 115-V primaries, dual 12-V, 2-A secondaries (Triad F107Z or equivalent)
(1) Resistor, 10-Ω, 25-W, ±5%
(1) Voltmeter, ac (VOM or DMM)
(1) Ammeter, ac clamp-on

Procedure

1. Use the VOM or DMM to measure the resistance of one of the 12-V windings of the transformer *before applying power* to the transformer. Record this resistance which is called the ohmic resistance of the winding. Later on you can compare the ohmic resistance to the internal resistance (r_t) of the transformer.

2. a. Construct the circuit in Fig. 23-9(a). Measure and record this voltage which we will identify as the full-load voltage, V_{FL}. *Note:* The 10-Ω resistor does not fully load the secondary, but is an adequate load for determining r_t of the transformer.
   b. Now, remove the 10-Ω resistor. Next, measure and record the open-circuit terminal voltage which we will identify as the no-load voltage, V_{NL}.

   c. The internal resistance of a transformer can be estimated by

   $$r_t = \frac{V_{NL} - V_{FL}}{I_{FL}}$$

   Since I_{FL} for our circuit is equal to V_{FL}/R_L, we can substitute for I_{FL} and write

   $$r_t = \frac{V_{NL} - V_{FL}}{V_{FL}/R_L} = \frac{(V_{NL} - V_{FL})R_L}{V_{FL}}$$

   With this formula, calculate the value of r_t using the data you have collected in the above steps.
   d. Is the internal resistance calculated in step 2c equal to the ohmic resistance measured in step 1? Why? (*Hint:* What causes a transformer to heat up when it is operating?)

3. a. Inspect Fig. 23-9(b). Notice that the jumper wire provides a dead short on the transformer secondary. Thus, this circuit cannot be powered for more than 2 or 3 s without overheating and perhaps destroying the transformer. Set up the circuit of Fig. 23-9(b) *but do not apply power.* Put the clamp-on ammeter on the 30-A range and be prepared to read the meter immediately after applying power to the circuit. Now apply power, read the meter, and remove the power. Record the short-circuit current just measured.
   b. Using the values of V_{NL} and r_t determined in step 2 and Ohm's law, you can estimate the short-circuit current. Calculate and record your

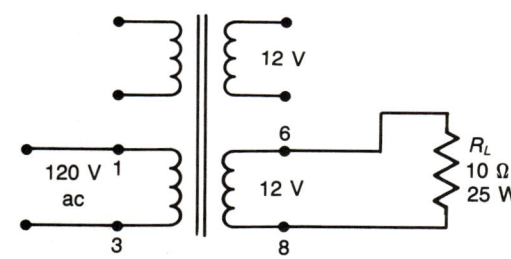

(a) Circuit for determining terminal voltage with heavy load

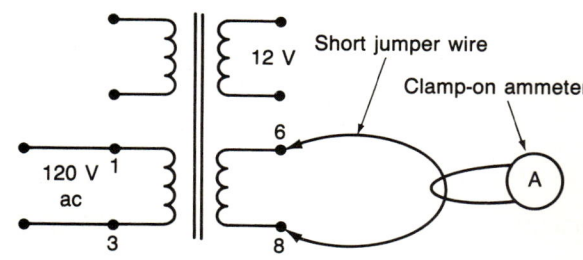

(b) Circuit for determining short-circuited current

FIGURE 23-9 Circuit for determining the internal resistance of a transformer.

estimate of the short-circuit current. Your estimate may be a little high because the short-circuit current may start to saturate the core material magnetically. How does this estimate compare with the value measured in step 3a?

4. Answer the following questions:
   a. A small transformer has an open-circuit voltage of 6.5 V and an internal resistance of 1.2 Ω. Determine the output voltage when the transformer is providing 0.3 A to a load.
   b. If the transformer in step 4a provided a peak recurrent current of 2 A in a filtered power supply, what would be the peak voltage out of the transformer?

ACTIVITY 23-4 FILTER CIRCUITS

Introduction

In this activity, you will connect a filter circuit to the output of a rectifier circuit and observe the effects of filtering. You will collect data that demonstrate the effects of R_L and C on ripple, output voltage, and peak repetitive rectifier current.

Supplies

(1) Transformer, dual 115-V primaries, dual 12-V, 2-A secondaries (Triad F 107Z or equivalent)
(1) Capacitor, 500-μF, 50 DCWV
(1) Capacitor, 1000-μF, 50 DCWV
(1) Resistor, 50-Ω, 25-W, ±5%
(1) Resistor, 100-Ω, 20-W, ±5%
(1) Resistor, 1000-Ω, 5-W, ±5%
(2) Diodes, IN4002
(1) Multimeter (VOM or DMM)
(1) Oscilloscope

Procedure

⬅ **Caution** The capacitors used in this activity may explode if operated with reverse polarity. Therefore, be certain you observe polarity on both the capacitor and the diode or diodes. Connect a voltmeter (with correct polarity) directly across the capacitor before powering the circuit. If the voltmeter does not immediately indicate the expected voltage when power is applied, immediately turn off the power. Correct any circuit errors before applying power again.

1. Half-wave filtered circuits
   a. In the circuits in Fig. 23-10, the common (ground) symbol indicates where the ground lead of the oscilloscope should be connected. The caption under each circuit lists the measurements to be made on the circuit. If you do not have a multimeter that measures the rms value of a nonsinusoidal waveform, then you cannot measure V_{rip}. In that case, you can estimate V_{rip} with the formula

   $$V_{rip} = 0.3 \times V_{rip,p\text{-}p}$$

   Measure $V_{rip,p\text{-}p}$, V_{R_1}, and PIV of D_1 with a calibrated ac-coupled oscilloscope. Measure the period of the ripple (T_{rip}) using the calibrated time base of the oscilloscope.
   b. Construct the circuit of Fig. 23-10(a) using a 500-μF, 50-V capacitor and a 50-Ω, 25-W resistor. Observing the CAUTION at the beginning of the procedure section, apply power to the circuit and make the measurements listed for this circuit. Record your measured values in the appropriate column of the table in Fig. 23-11.
   c. Change R_L in the circuit of Fig. 23-10(a) to a 100-Ω, 20-W resistor. Again make the listed measurements and record the measured values in the appropriate column in Fig. 23-11. What happened to the output voltage and the ripple when R_L was changed from 50 to 100 Ω?
   d. Next change C_1 to 1000 μF; double-check the polarity of C_1 and use the procedure discussed in the CAUTION statement. Make the required measurements and record them in the table in Fig. 23-11.
   e. Now change R_L back to 50 Ω. Make the required measurements and record them in the table.
   f. Change C_1 back to 500 μF and add R_1 so that you have the circuit shown in Fig. 23-10(b). Check the polarity of C_1, then measure and record the values of V_{R_1} and PIV.
   g. Finally, change R_L back to 100 Ω and again measure and record the value of V_{R_1}. This completes the measurements of the half-wave filtered circuits.
   h. With the data you have collected, you can now do the calculations needed to complete the H-W columns in the table of Fig. 23-11. The appropriate formulas are:

   $$\text{rf} = \frac{V_{rip}}{V_{dc}}$$

   $$I_{fpr} = \frac{V_{R_1}}{R_1}$$

   $$f_{rip} = \frac{1}{T_{rip}}$$

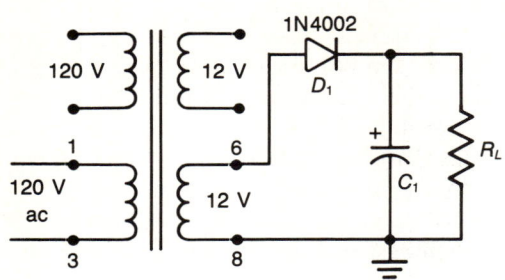
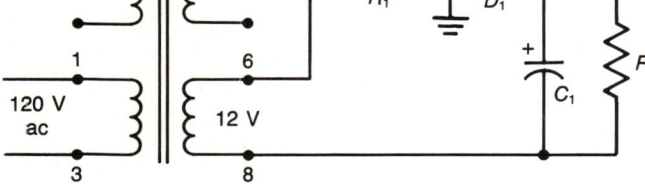

(a) Half-wave circuit for measuring V_{ac}, $V_{(rip)\ P\text{-}P}$, $V_{(rip)}$ and $T_{(rip)}$

(b) Half-wave circuit for measuring V_{R1} and PIV of D_1

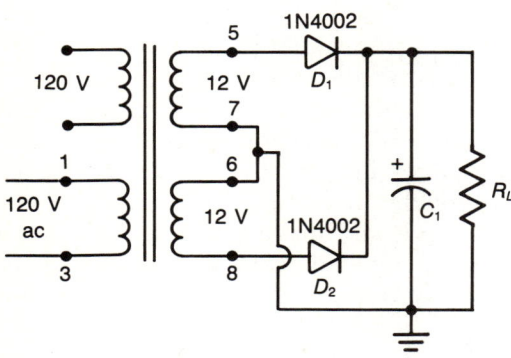
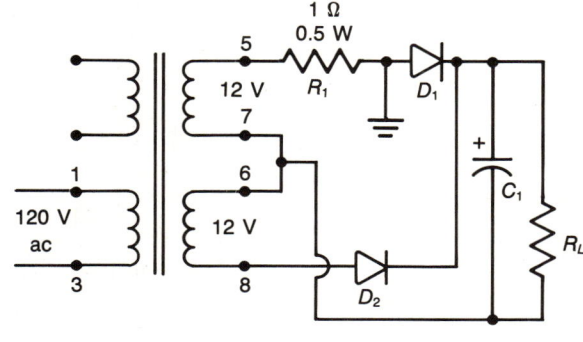

(c) Full-wave circuit for measuring V_{ac}, $V_{(rip)\ P\text{-}P}$, $V_{(rip)}$ and $T_{(rip)}$

(d) Full-wave circuit for measuring V_{R1} and PIV of D_1

FIGURE 23-10 Circuits for determining filter characteristics.

Complete the H-W columns at this time.

i. Using the data in the H-W columns in the table in Fig. 23-11, answer these questions:
- If C_1 is doubled and R_L is halved, what happens to rf?
- Does I_{fpr} increase or decrease when the load current increases?
- How much load current is drawn by the 50-Ω resistor when C_1 is 1000 μF?

2. Full-wave filtered circuits
   a. You are now ready to construct the circuits in Fig. 23-10(c) and (d) and make the measurements called for in the F-W columns in Fig. 23-11. Repeat steps 1b through h but with these changes: H-W becomes F-W, Fig. 23-10(a) becomes Fig. 23-10(c), and Fig. 23-10(b) becomes Fig. 23-10(d).
   b. Using the formulas of step 1h, complete the F-W columns of Fig. 23-11.
   c. Does the full-wave filtered circuit have a higher value of I_{fpr}/I_{dc} when R_L is 50 Ω or when R_L is 100 Ω?

3. Voltage regulation
   a. Reconstruct the circuit of Fig. 23-10(c) using a 500-μF capacitor and a 1000-Ω, 5-W resistor. Consider the 1000-Ω resistor to be a bleeder resistor. Measure and record V_{dc}, which will be V_{NL} for figuring the percentage of voltage regulation. Now parallel the 1000-Ω resistor with a 50-Ω, 25-W resistor and again measure V_{dc}. Record this value of V_{dc} as V_{FL}.
   b. In case you have forgotten, the formula for percentage of voltage regulation is

   $$\% \text{ of } VR = \frac{V_{NL} - V_{FL}}{V_{FL}} \times 100$$

   Using the data of step 3a, calculate and record the percent of VR.

4. As a review-summary-comparison of filtered power-supply circuits, answer these questions. (Refer to Fig. 23-11 for help if necessary.) All other factors (C, R_L, etc.) being equal, which circuit (H-W or F-W) provides the:
   a. least ripple
   b. least I_{fpr}
   c. lowest f_{rip}
   d. most V_{dc}

	$C_1 = 500\ \mu F$				$C_1 = 1000\ \mu F$			
	$R_L = 50\ \Omega$		$R_L = 100\ \Omega$		$R_L = 50\ \Omega$		$R_L = 100\ \Omega$	
	H-W	F-W	H-W	F-W	H-W	F-W	H-W	F-W
V_{dc}								
$V_{(rip)p\text{-}p}$								
$V_{(rip)}$								
rf								
V_{R_1}								
I_{fpr}								
$T_{(rip)}$								
$f_{(rip)}$								
PIV								

FIGURE 23-11 Data table for circuits of Fig. 23-10.

ACTIVITY 23-5 ZENER REGULATORS

Introduction

This activity demonstrates the voltage regulation capability of a zener diode circuit. It also shows how a zener circuit reduces ripple.

Supplies

(1) Transformer, dual 117-V primaries, dual 12-V, 2-A secondaries (Triad F107Z or equivalent)
(1) Diode, rectifier-type, IN4002
(1) Diode, zener, 5.6-V ±5%, 1-W, IN4733B
(1) Resistor, 100-Ω, 1-W, ±5%
(1) Resistor, 100-Ω, 20-W, ±5%
(1) Resistor, 125-Ω, 3-W, ±5%
(1) Capacitor, 1000-μF, 50-DCWV
(1) Multimeter (VOM or DMM)
(1) Oscilloscope

Procedure

↪ **Caution** The capacitors used in this activity may explode if operated with reverse polarity. Therefore, be certain you observe polarity on both the capacitor and the diode or diodes. Connect a voltmeter (with correct polarity) directly across the capacitor before powering the circuit. If the voltmeter does not immediately indicate the expected voltage when power is applied, immediately turn off the power. Correct any circuit errors before applying power again.

1. The half-wave power supply in Fig. 23-12 produces an unregulated output voltage across R_1 and the zener regulator. The zener regulator circuit produces a smaller, but well-regulated, output voltage across R_L. For this circuit, R_L can be any value that is greater than 95 Ω.
2. Ripple voltage reduction
   a. Construct the circuit of Fig. 23-12, apply power and measure the ripple voltage V_{rip} across R_1 with either an rms-reading DMM or an oscilloscope. If you measure the ripple with an oscilloscope, multiply the peak-to-peak value by 0.3 to get an estimate of the rms value. Record the value of V_{rip} for later use.
   b. Measure and record V_{dc} across R_1.
   c. Using the values obtained in steps 2a and b, calculate and record the ripple factor of the voltage across R_1. This is the ripple factor of the voltage input to the zener regulator. The formula is rf = V_{rip}/V_{dc}.
   d. Next, measure and record V_{rip} and V_{dc} across R_L. Now calculate and record the ripple factor for the voltage output from the zener regulator.
   e. Compare the ripple factors from steps 2c and d. By what factor did the zener regulator reduce the ripple factor?
3. Regulation
   a. The zener regulator is fully loaded by R_L. Therefore, you have already measured V_{FL} in step 2d. Now remove R_L from the circuit; now measure and record the voltage across the zener diode. This is V_{NL}.
   b. With the data you now have, calculate and record the percent of voltage regulation for the zener regulator circuit. The formula is

$$\% \text{ of } VR = 100 \times \frac{V_{NL} - V_{FL}}{V_{FL}}$$

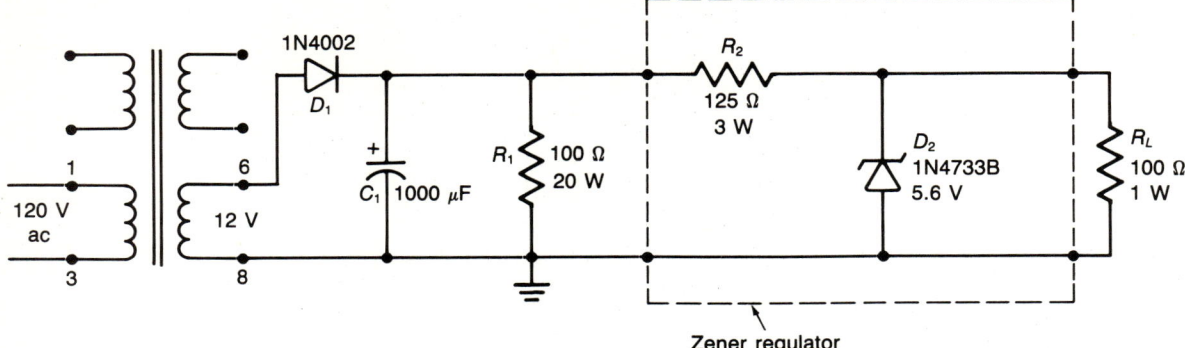

FIGURE 23-12 Circuit for determining the characteristics of a zener regulator.

4. Dynamic resistance
   a. The dynamic resistance of the zener (r_z) can be estimated with the formula $r_z = \Delta V_z / \Delta I_z$. You already have the data for figuring ΔV_z because $\Delta V_z = V_{NL} - V_{FL}$. All you need now is the value of the zener current under no-load and full-load conditions.
   b. While R_L is still removed, measure the zener current and record it as $I_{z,NL}$. *Remember:* Interrupt the circuit and insert the ammeter in series with the zener.
   c. Next, reconnect R_L and again measure the zener current. Record this current as $I_{z,FL}$.
   d. Now you can calculate r_z with the following formula:

   $$r_z = \frac{V_{NL} - V_{FL}}{I_{z,NL} - I_{z,FL}}$$

   Record your calculated value for r_z for later use.
5. Check your understanding of zener regulators by answering the following questions:
   a. If the current through D_2 in Fig. 23-12 changed from 10 to 150 mA, how much would the voltage across D_2 change? (Use your value of r_z from step 4d.)
   b. When the load current (I_{R_L}) increases by 20 mA, what happens to the zener current?
   c. When the load current decreases by 30 mA, what happens to the current through R_2?

ACTIVITY 23-6 LOW-VOLTAGE POWER SUPPLY

Introduction

The no-load dc output of a filtered power supply is primarily dependent on the voltage of the transformer secondaries and how the secondaries are configured. The loaded dc output voltage varies with the magnitude of the load and the size of the filter capacitor.

Supplies

(1) Transformer, dual 115-V primaries, dual 12-V, 2-A secondaries (Triad F 107Z or equivalent)
(1) DMM
(1) Oscilloscope

Miscellaneous capacitors, resistors and diodes selected from the Materials List in this manual.

Procedure

↪ **Caution:** The capacitors used in this activity may explode if operated with reverse polarity. Therefore, be certain that you observe polarity on both the capacitor and the diode or diodes. Connect a voltmeter (with correct polarity) directly across the capacitor before powering the circuit. If the voltmeter does not immediately indicate the expected voltage when power is applied, immediately turn off the power. Correct any circuit errors before applying power again.

1. Design a filtered power supply that will provide a no-load dc output of approximately 34 V and have a ripple frequency of 120 Hz.
2. Construct and test your design. Use several different loads.
3. Write a report for this activity. Follow the guidelines in Appendix 1.

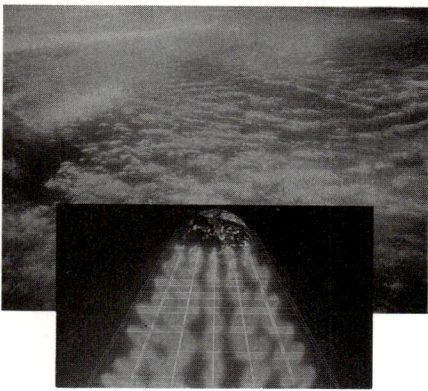

Chapter 24

Electronic Amplification

ACTIVITY 24-1 TRANSISTOR CHARACTERISTICS

Introduction

The similarities and differences between bipolar transistors (BJTs) and junction field-effect transistors (JFETs) will be investigated in this activity. As you progress through the activity, you will learn how to use an ohmmeter to determine the condition of a transistor and how to determine whether a JFET is N-channel or P-channel or a BJT is NPN or PNP.

Supplies

- (1) Power supply, 0- to 25-V dc
- (1) Battery, 6-V; or another dc power supply
- (1) JFET, 2N5458 or equivalent
- (1) BJT, 2N4124 or equivalent
- (1) BJT, 2N4126 or equivalent
- (1) Potentiometer, 1-kΩ, ½-W, ±5%, linear
- (1) Resistor, 22-kΩ, ½-W, ±5%
- (1) Capacitor, 500-µF, 50 DCWV
- (1) Multimeter, 10-MΩ input resistance (DMM or electronic MM)
- (1) Ammeter, with microampere range (DMM or VOM)

Procedure

1. Look over the BJT (2N4124) and see if the leads are identified by the letters E, B, and C printed on the case. Some manufacturers identify the leads with printed letters, others don't. If the leads are not identified, refer to Fig. 24-1(a). In this figure the leads are identified in a bottom view of the transistor. This is the standard way of identifying leads; however, to aid in lead identification, a pictorial drawing of the transistor is also shown in Fig. 24-1(b). The lead identification for the 2N5458 JFET is shown in Fig. 24-1(c).

   a. Select the $R \times 10$ range of the multimeter (or the resistance range of the DMM indicated for checking diodes). Connect the negative (−) lead of the ohmmeter to the base and the positive (+) lead to the collector of the 2N4124 which is an NPN transistor. This will reverse-bias the B-C junction so the resistance should be very high. *Do not* switch to the highest resistance range of the VOM because the internal voltage source of the VOM may exceed the reverse-voltage rating of the transistor's junctions (BV_{CBO} or BV_{BEO}). Move the + lead from the collector to the emitter. Is the base-emitter junction forward-biased or reverse-biased? Why?

(a) Lead identification for the 2N4124 (bottom view)

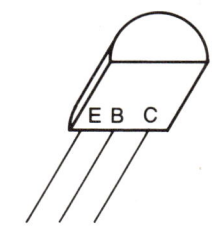

(b) Pictorial of the 2N4124 with leads identified

(c) Lead identification for the 2N5458 (bottom view)

FIGURE 24-1 Transistor lead identification.

b. If the same ohmmeter connections as above were used with a PNP, would the resistance be high or low? Why?

c. If you connect the + lead to the base and the − lead to the collector of the 2N4124, will the ohmmeter indicate a high or a low resistance? Why? Check your answer.

d. Now move the − lead from the collector to the emitter of the 2N4124. Is the resistance about the same as when the − lead was on the collector? Why?

e. Next check the resistance from emitter to collector using first one polarity and then the other. Is the resistance high or low in both directions? Why?

f. Notice from the measurements in steps 2a and d that the resistance from the base lead to either of the other two leads changes from high to low or low to high when the polarity is reversed. This is true for the base lead only. Using ohmmeter measurements, determine and record which lead of the 2N4126 is the base lead. Is the 2N4126 a PNP or an NPN? Why?

2. Checking JFETs

a. The 2N5458 is a JFET with the lead configuration shown in Fig. 24-1(c). Connect the + lead of the ohmmeter to the gate and the − lead to the source. Is the resistance high or low? Does the 2N5458 have a P channel or an N channel? Why?

b. Will moving the − lead to the drain result in a high or low resistance? Why? Check your answer.

c. Put the − lead of the ohmmeter on the gate and the + lead to the source and then to the drain. Describe the results of these two measurements.

d. Now measure the resistance from source to drain. Does the polarity of ohmmeter leads affect this resistance? Is the resistance as large as the resistance of the reverse-biased gate-to-channel resistance? Why?

e. Is the source-to-drain resistance of the JFET as large as the emitter-to-collector resistance of the BJT? Why?

3. Characteristics of the BJT

a. Refer to Fig. 24-2, which is a circuit you can use to determine the behavior of an NPN transistor with various values of I_B. The 22-kΩ resistor limits I_B to a safe level even if R_1 is accidentally turned so that the full 6 V of B_1 is applied to the base circuit. R_1 is included in the circuit so that fine voltage adjustments (and thus I_B adjustments) can be made. The 500-μF capacitor keeps the circuit from oscillating. It does not affect the dc conditions. Construct the circuit shown in Fig. 24-2 and adjust the center arm of R_1 toward its common end. Then, when power is applied, the base current will be small. When working with this circuit, it is best to leave the current meter A_1 in the circuit at all times. (A current meter may have so much internal resistance that removing it would cause a significant circuit resistance change, which would cause a significant charge in I_B).

In working with the circuit in Fig. 24-2, you may find that the collector current keeps creeping higher and higher if the circuit is left on for an extended period of time. This is caused by a temperature increase in the transistor as the collector dissipates heat. The collector current creep is most noticeable when I_C and V_C are both at their highest values. To minimize this effect, keep the base current at zero except while reading the collector current meter or measuring the base-emitter voltage.

b. Recheck the polarity of C_1, then apply power to the circuit of Fig. 24-2 and adjust R_1 until I_B indicates zero. Read the collector current. Record this current in the upper-left-hand cell of the table in Fig. 24-3. Adjust R_1 for an I_B of 50 μA. Read the collector current meter and then reduce I_B back to zero so the transistor can stay cool. Record the value of I_C when I_B = 50 μA in the appropriate cell of the data table.

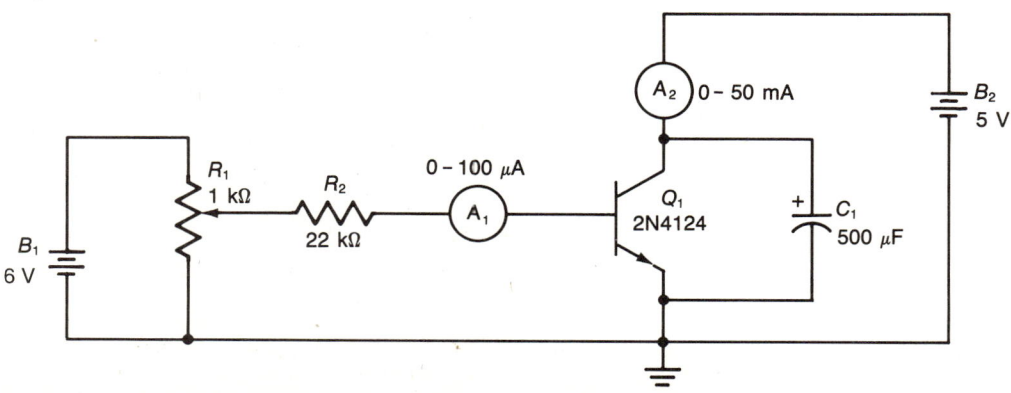

FIGURE 24-2 Circuit for measuring the characteristics of an NPN transistor.

V_{CE}	I_C at $I_B = 0\ \mu A$	I_C at $I_B = 50\ \mu A$	I_C at $I_B = 100\ \mu A$
5 V			
10 V			
15 V			

FIGURE 24-3 Data table for the circuit of Fig. 24-2.

Connect a high-input-impedance voltmeter between the base and emitter of Q_1, then increase I_B to 50 μA and read V_{EB}. Reduce I_B back to zero and record the value of V_{EB}. Now increase I_B to 100 μA, measure V_{EB} and reduce I_B to zero. Record this value of V_{EB}. Again increase I_B to 100 μA, read the I_C meter and reduce I_B to zero. Record I_C in the appropriate cell of Fig. 24-3. How much did V_{EB} increase when I_B increased from 50 to 100 μA? How much increase in I_C was caused by this 50-μA increase in I_B?

c. Remove the voltmeter from the B-E and increase B_2 to 10 V. Using the procedures followed in step 4b above, make the three I_C measurements needed to complete the second row of Fig. 24-3. Does doubling the collector-to-emitter voltage increase I_C as much as doubling I_B? Why?

d. Increase B_2 to 15 V and make the measurements needed to complete the last row in Fig. 24-3.

e. Using the data from Fig. 24-3, calculate and record β at $V_{CE} = 5$ V and $I_B = 50\ \mu A$. Also calculate and record β at $V_{CE} = 15$ V and $I_B = 100\ \mu A$. Which of these calculated β's is higher?

f. The manufacturer of the 2N4124 lists the dc β as 120 minimum. Do your calculated β's exceed this value? Compare your calculated β's with those of other class members. Do they agree within 20 percent?

g. Was Q_1 in Fig. 24-2 ever saturated? Was it ever cut off?

4. Characteristics of the JFET
   a. Construct the circuit of Fig. 24-4 but leave the voltmeter out of the circuit. *Notice that the polarity of B_1 is reversed from what it was in Fig. 24-2.* Insert a current meter (low microampere range) between the gate lead and R_1. Turn on the power and turn R_1 through its entire range while observing the meter in the gate lead. Does the gate draw any measurable current? Why?
   b. Turn off the power and remove the current meter from the gate lead. Connect a high-input-impedance voltmeter between gate and source as shown in Fig. 24-4. Adjust R_1 so that the center arm is at the negative end of R_1. Apply power and read the drain current indicated by

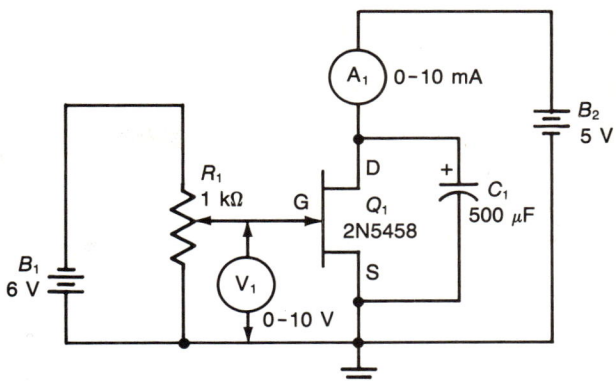

FIGURE 24-4 Circuit for measuring the characteristics of an N-channel JFET.

V_{DS}	I_D at $V_{GS} = 0$ V	I_D at $V_{GS} = 0.5$ V	I_D at $V_{GS} = 1.0$ V
5 V			
10 V			
15 V			

FIGURE 24-5 Data table for the circuit of Fig. 24-4.

A_1. Is the transistor in cutoff or saturation? Compare the results of applying a negative voltage to the gate of an N-channel JFET to the results of applying a positive voltage to the base of an NPN BJT.

c. Reduce V_{GS} to zero and read the drain current meter; then, increase V_{GS} back to 6 V. Record the I_D just measured in the upper-left-hand cell of the table in Fig. 24-5. Change V_{GS} to 0.5 V, read I_D, and then change V_{GS} back to 6 V. Record I_D in Fig. 24-5. Continue this procedure until the table in Fig. 24-5 is completed; remember to keep V_{GS} at 6 V except when reading I_D.

d. For the 2N5458 JFET, the manufacturer specifies I_{DSS} at 2 to 9 mA when $V_{DS} = 15$ V and $V_{GS} = 0$ V. (I_{DSS} is the symbol for drain current when the gate is shorted to source.) Does the value of I_{DSS} you measured (lower-left-hand cell of Fig. 24-5) fall between 2 and 9 mA?

e. The manufacturer of the 2N5458 JFET specifies that the drain current will be cut off (less than 10 nA) by a gate-to-source voltage of between 1 and 7 V when V_{DS} is 15 V. The gate voltage needed to cut off drain current is called $V_{GS,off}$. Using the circuit of Fig. 24-4, determine and record the value of $V_{GS,off}$ for your JFET. Is it within the range specified?

f. Why is a JFET referred to as a voltage-controlled device?

g. Does a 50 percent change in V_{DS} have as much effect on I_D as a 50 percent change in V_{GS} when the JFET is not cut off?

ACTIVITY 24-2 TRANSISTOR CURVES

Introduction

This activity will show you how to display on an oscilloscope the drain and collector characteristic curves for FETs and BJTs, respectively. You will also learn how to calibrate and use the external input to the horizontal amplifier in the oscilloscope. In this activity, you will display curves for a BJT, a JFET, and a MOSFET.

Supplies

- (2) Batteries, 6-V; or, dc power supplies with ungrounded output
- (1) BJT, 2N4124 or equivalent
- (1) JFET, 2N5424 or equivalent
- (1) MOSFET, 2N3797 or equivalent
- (1) Transformer, dual 115-V primaries, dual 12-V, 2-A secondaries (Triad F107Z or equivalent); or any 12-V ac source with an ungrounded output
- (1) Diode, 1N4002 or equivalent
- (1) Potentiometer, 1-kΩ, $\frac{1}{2}$-W, ±5%, linear
- (1) Resistor, 10-Ω, $\frac{1}{2}$-W, ±5%
- (1) Resistor, 100-Ω, $\frac{1}{2}$-W, ±5%
- (2) Resistors, 1-kΩ, $\frac{1}{2}$-W, ±5%
- (1) Resistor, 22-kΩ, $\frac{1}{2}$-W, ±5%
- (1) Capacitor, 0.01-μF, 25 DCWV
- (1) Multimeter, with microampere range (DMM or VOM)
- (1) Oscilloscope, 10-mV/cm vertical sensitivity and external input to the horizontal amplifier

Procedure

1. BJT curves
   a. The circuit for displaying the collector characteristic curve of a BJT is shown in Fig. 24-6. The base circuit is identical to that used in the previous activity except that C_1 has been added to prevent the circuit from oscillating. R_1 controls the base current. In the collector-to-emitter circuit, T_1 and D_1 form a half-wave rectifier that provides a pulsating dc voltage for the collector. R_3, being in series with the collector supply voltage, monitors the collector current and provides a voltage which is directly proportional to the collector current. The voltage across R_3 goes to the vertical input of the oscilloscope. Thus, the vertical deflection on the scope represents collector current. The value of collector current at any point on the displayed curve can be determined by Ohm's law: $I_C = V_{R3}/R_3$. For example, suppose the curve in Fig. 24-7 was displayed on the scope and the vertical channel was calibrated for 20 mV/div. Then the vertical voltage at the intersection of 4 vertical divisions and 9 horizontal divisions would be 20 mV/div. × 4 div. = 80 mV. This would represent a collector current of

   $$I_C = \frac{V_{R3}}{R_3} = \frac{80 \text{ mV}}{10 \text{ }\Omega} = 8 \text{ mA}$$

   The voltage into the horizontal amplifier of the oscilloscope is V_{CE}. If the horizontal (X) axis is calibrated for 2 V/div., then the ninth division in Fig. 24-7 would represent an instantaneous V_{CE} of 2 V/div. × 9 div. = 18 V. Thus, the graph in Fig. 24-7 tells us that the instantaneous I_C is 8 mA when the instantaneous V_{CE} is 18 V. With the same calibrations as used above, what is the instantaneous value of I_C for the curve in Fig. 24-7 at 4 horizontal divisions?

   b. Before displaying a curve, you must calibrate the horizontal channel of the oscilloscope using the following procedure. First, connect the *vertical input* to a low-voltage ac source such as a 12-V secondary of a transformer or a signal generator and obtain about a 1-in vertical deflection. This will keep the electron beam from being stationary and burning the CRT phosphor when horizontal sweep is removed. Second, adjust a variable dc power supply for 20.0 V and connect the negative terminal of the power supply to the scope ground. Third, turn the horizontal-time-base-selector switch on the oscilloscope to the "External H" or "External X" position and notice that the scope displays a vertical line. Now, connect the positive lead of the power supply to the "H input," or "X input," and observe which way the line deflects. Remove the positive lead of the power supply and, using the X axis position control, position the vertical line to the last graticule mark on the side opposite to the direction the line deflected. Reconnect the positive lead of the power supply to the X input jack and adjust the X axis gain control so that the vertical line moves 10 div. when the +20-V lead is reconnected. Finally, disconnect the positive lead of the power supply and see if the vertical line returns 10 div.; if not, repeat the procedure until it does. The horizontal channel (X axis) is now calibrated for 2 V/div.; *do not change the X axis gain or position controls* for the rest of this activity. If the vertical line is deflected 3.8 div. horizontally, what is the input voltage to the X input jack?

   c. Next, the electron beam has to be positioned to either the top or bottom of the screen so that a

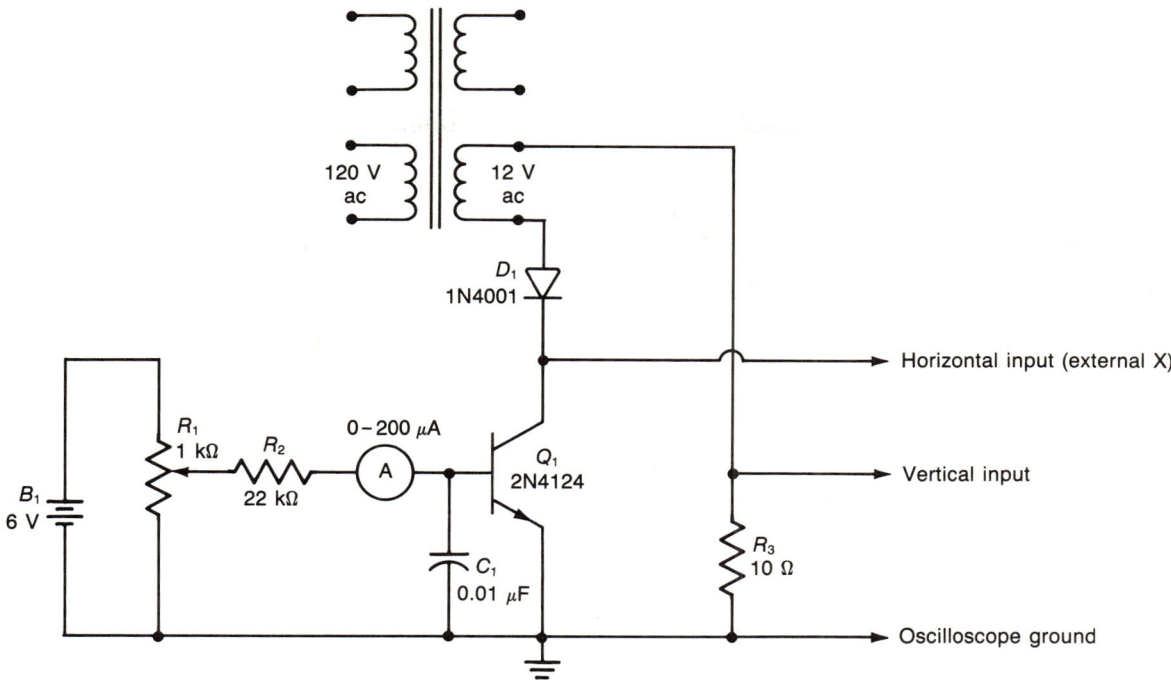

FIGURE 24-6 Circuit for displaying the collector characteristic curve of an NPN transistor.

negative-going vertical input will deflect the beam toward the center of the screen. The steps are:

- Connect the X input jack to the same ac source as the vertical input is connected. *Then,* remove the vertical input from the ac source and notice that a horizontal line is displayed.
- Set the vertical calibration controls for 500 mV/div.
- Now remove the − lead of the dc power supply from the scope ground and connect the + lead to the scope ground. Adjust the power supply for about 1 V. Set the vertical input to "dc-coupled" and momentarily touch the − lead of the power supply to the vertical input while observing which way the horizontal line deflects. Use the vertical position control to position the beam to the top or bottom graticule line as indicated by the direction of the deflection.

d. The electron beam should now be positioned in one of the corners of the graticule grid. To check if it is, turn the intensity or brightness control fully counterclockwise. The horizontal line should completely disappear and nothing should be seen on the screen. Now disconnect the X axis input from the low-voltage source. Connect both the vertical input and the X axis input to the scope ground. This will keep stray induced voltages from entering the amplifiers. Now slowly increase the intensity control until a *faint* dot appears on the screen; then *immediately* reduce the intensity control to remove the dot. Leave the intensity control in this position until you are ready to observe a characteristic curve.

e. Construct the circuit shown in Fig. 24-6. Adjust R_1 to its ground end. Then turn on both the 6-V dc supply and the 12-V ac supply. Turn up the intensity control on the oscilloscope and observe the horizontal line. This line is the collector curve for $I_B = 0$. What is the maximum value of V_{CE} shown on the scope? What is the maximum value of I_C shown?

f. Turn R_1 until the $I_B = 150$ μA and observe the collector curve. What is the maximum value of I_C and V_{CE} now? Carefully plot this curve on a graph like that in Fig. 24-8. Accurately label and mark the divisions on both the horizontal and vertical axis of the graph.

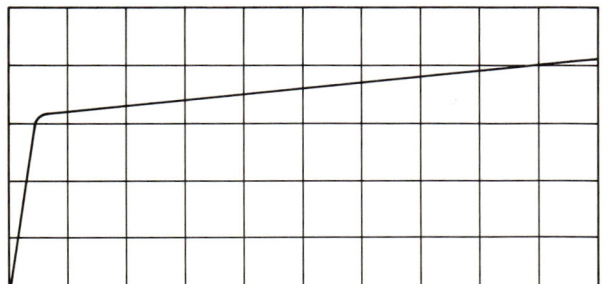

FIGURE 24-7 Typical collector curve displayed on an oscilloscope. (The quadrant in which the curve appears depends upon the oscilloscope used.)

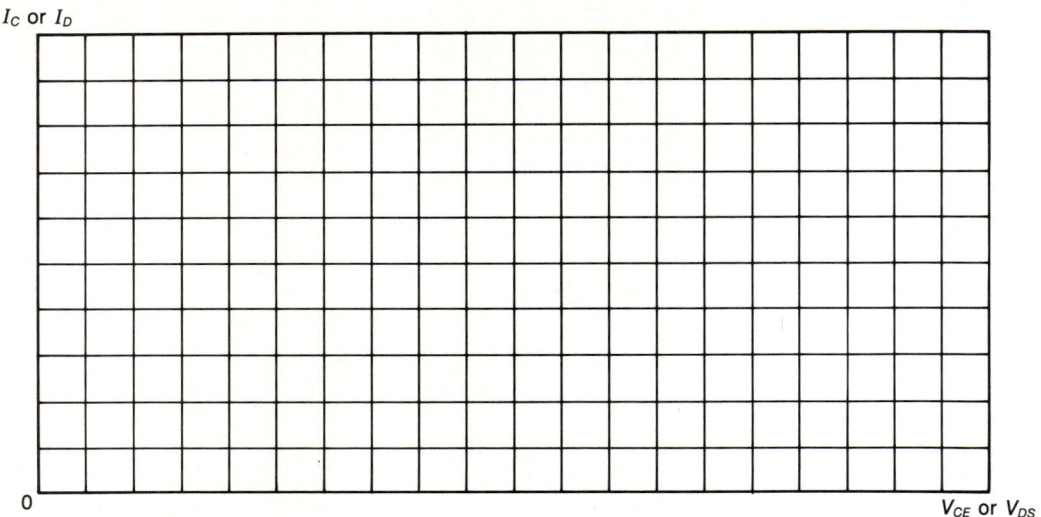

FIGURE 24-8 Sample of graph on which to plot curves.

g. Reduce I_B to 100 µA, record the maximum I_C, and draw the curve on the same graph paper as used in step 1f.

h. Reduce I_B to 50 µA and once again draw the observed curve on the same graph with the previous two curves.

i. From the curves you have drawn, calculate and record β at $I_B = 100$ µA and $V_{CE} = 10$ V.

j. Reduce the intensity control setting on the scope until the trace disappears. Turn off both the ac and dc power supplies.

2. JFET curves
   a. Construct the circuit shown in Fig. 24-9. Note the polarity of B_1 and the increased value of R_2. For maximum drain current, should R_1 be adjusted toward ground or toward the negative end of B_1?
   b. Adjust R_1 for maximum drain current, turn on both power supplies, and then increase the intensity control of the scope to observe the drain characteristic curve with $V_{GS} = 0$. What is the value of I_{DSS} when $V_{DS} = 15$ V?

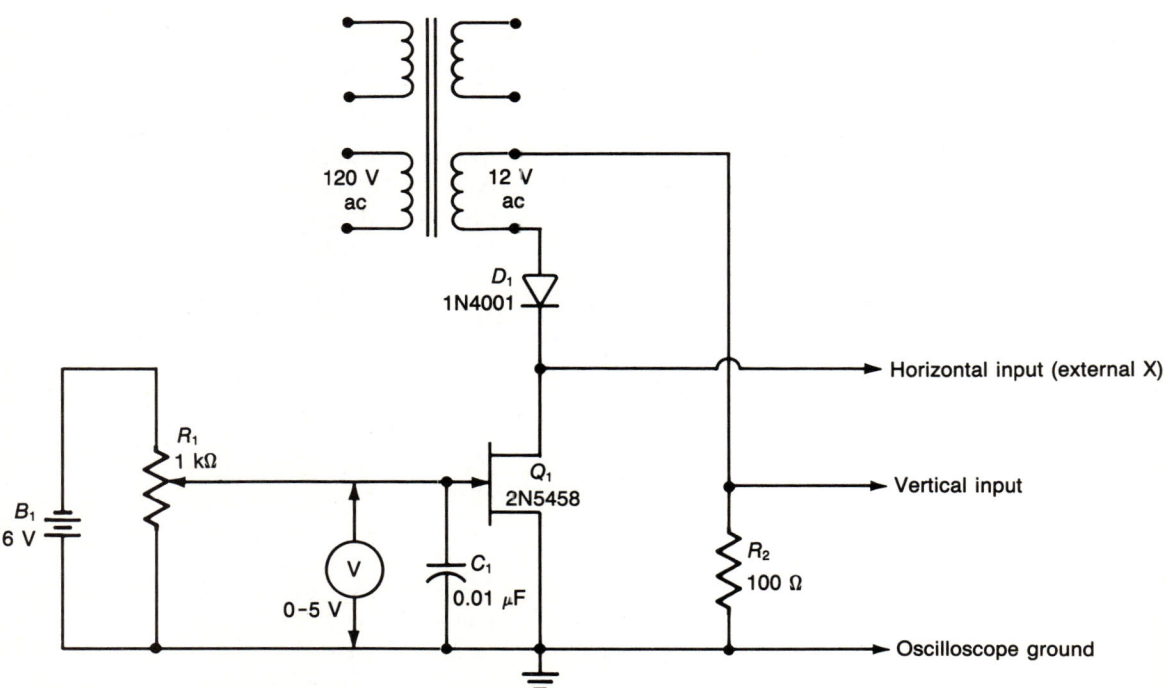

FIGURE 24-9 Circuit for displaying the drain characteristic curve of an N-channel JFET.

c. On a graph like the one in Fig. 24-8, accurately draw the curves you observe when V_G is 0, -0.5 V, -1.0 V, and -1.5 V.
d. Are the curves for the JFET as evenly spaced as they are for the BJT?
e. Reduce the intensity control on the oscilloscope and then turn off the ac and dc power supplies.
3. MOSFET curves
   a. MOSFETs are easily damaged by static charge. Leave the MOSFET in its static-protected package until the rest of the circuit is constructed and ready for power. Before removing the static protection from the MOSFET, short its leads together by wrapping a very fine wire around the leads just below the case of the MOSFET. Hold the 2N3797 by the case when installing it in the circuit. Construct the circuit in Fig. 24-10(a); install the 2N3797 in the circuit before removing the shorting wire from its lead. Now turn on all power supplies, and adjust R_1 toward the end connected to R_2. This will make the gate about $+3$ V with respect to the source. Will the MOSFET be operating in the depletion or enhancement mode?
   b. Turn up the intensity on the scope and observe the drain curve. What is the value of I_D at $V_{DS} = 15$ V?
   c. On a graph like those used for the BJT and the JFET, draw the curves you observe when V_G is $+1.0$ V, 0 V, and -1.0 V.
d. Turn the horizontal selector switch on the oscilloscope to any internal sweep position. Turn off all power supplies. Short the MOSFET leads with a fine wire; then remove the MOSFET and put it back in its static-protected package.
e. What is the value of I_{DSS} at $V_{DS} = 10$ V for the 2N3797 that you are using?
f. Are the curves of the MOSFET as evenly spaced as the curves of the JFET?

ACTIVITY 24-3 BJT AMPLIFIERS

Introduction

The CC and the CE amplifiers will be constructed and tested in this activity. The effects of negative feedback on voltage gain will be observed and the input impedance of an amplifier will be indirectly measured.

Supplies

(1) Power supply, 20-V dc, floating output
(1) BJT, 2N4124 or equivalent
(1) BJT, 2N4126 or equivalent
(1) Potentiometer, 50-kΩ, $\frac{1}{2}$-W, $\pm 5\%$, linear
(1) Resistor, 560-Ω, $\frac{1}{2}$-W, $\pm 5\%$

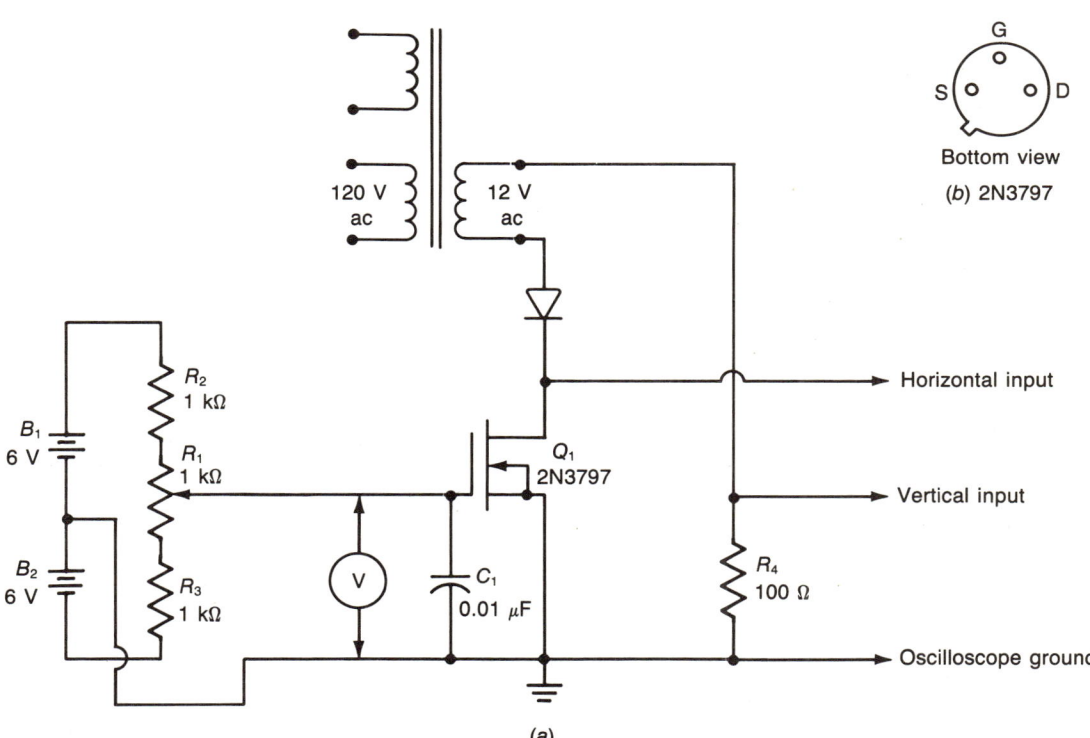

FIGURE 24-10 Circuit for displaying the drain characteristic curve of an N-channel MOSFET.

- (2) Resistors, 1-kΩ, ½-W, ±5%
- (1) Resistor, 5.1-kΩ, ½-W, ±5%
- (2) Resistors, 10-kΩ, ½-W, ±5%
- (1) Resistor, 22-kΩ, ½-W, ±5%
- (2) Resistors, 68-kΩ, ½-W, ±5%
- (1) Resistor, 180-kΩ, ½-W, ±5%
- (2) Capacitors, 10-μF, 25 DCWV
- (1) Capacitor, 500-μF, 50 DCWV
- (1) Signal generator, audio range
- (1) Oscilloscope, ×10 probe, external trigger
- (1) Multimeter (DMM or VOM)

Procedure

1. CE amplifier
   a. For Fig. 24-11, calculate the quiescent quantities listed in Fig. 24-12(a). Enter the calculated values in the first column of the table.
   b. Construct the circuit in Fig. 24-11, but omit C_E and R_L. Recheck the polarity of C_i and C_o. (*Caution*: Reverse polarity can cause a capacitor to explode.) Turn on the V_{CC} supply but leave the signal generator off. Measure, and record in the second column of Fig. 24-12(a), the quiescent quantities calculated above. How closely do the measured and calculated values agree? If they are not within 15 percent, check both again.
   c. What should be the phase relationship of the input and output signals in Fig. 24-11?
   d. Set the oscilloscope for external trigger and connect a lead from the − end of C_o to the external-trigger input jack. Connect the vertical input lead (ac coupling) to this same point. Now turn on the signal generator and adjust its output until the oscilloscope shows a 2 $V_{p\text{-}p}$ output signal. Notice where on the screen the signal reaches its peak positive value. Now move the vertical input lead to the negative end of C_i and notice where the input signal goes positive. Was your answer to step 1c correct? Remove the external-trigger lead and set the oscilloscope for internal triggering.
   e. Using the circuit values given in Fig. 24-11, calculate, and record in Fig. 24-12(b), the voltage gain when R_L and C_E are removed. Next, using the oscilloscope to measure the input and output voltages, determine the measured voltage gain and record it in Fig. 24-12(b). In measuring voltage gain, the signal generator output level control should be adjusted so that the output signal is about 50 percent of the value where clipping of the signal is first noticed. Is the voltage gain you calculated within 15 percent of the gain you measured? It should be!
   f. Adding the values of R_L and bypassing the indicated amount of R_E, complete the table in Fig. 24-12(b). *Observe polarity on the bypass capacitor.*

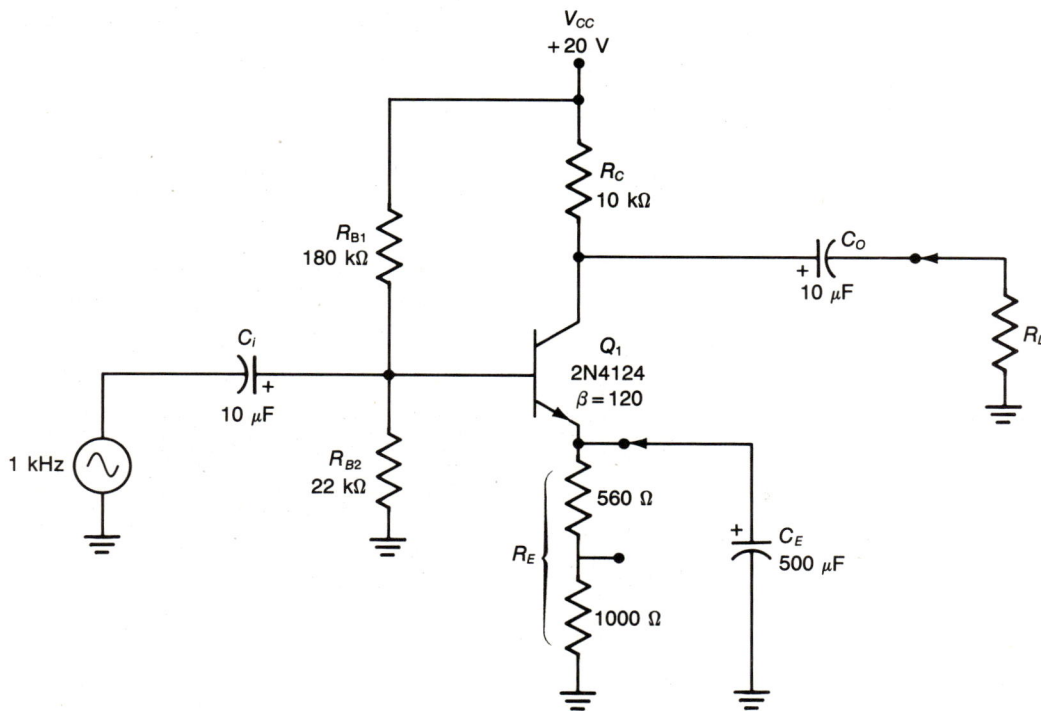

FIGURE 24-11 CE amplifier.

| | Quiescent values ||
	Calculated	Measured
V_B		
V_E		
V_C		
V_{CE}		
I_C		

(a)

| Unbypassed R_E | Voltage gain ||||||
| | No-load || $R_L = 10\ k\Omega$ || $R_L = 1\ k\Omega$ ||
	Calc.	Meas.	Calc.	Meas.	Calc.	Meas.
1560 Ω						
560 Ω						
0 Ω						

(b)

| Unbypassed R_E | Input impedance ||
	Calculated	Measured
1560 Ω		
560 Ω		
0 Ω		

(c)

FIGURE 24-12 Data tables for the circuit in Fig. 24-11.

g. Under what condition did the CE amplifier have the largest voltage gain?
h. Under what condition did the CE amplifier have the smallest voltage gain?
i. Under what condition was the calculated voltage gain least accurate? Why?
j. Next, calculate, and record in Fig. 24-12(c), the Z_i for the circuit of Fig. 24-11 when various amounts of R_E are bypassed. Next, measure and record Z_i by inserting a 50-kΩ potentiometer, connected as a rheostat, between the output of the signal generator and the negative end of capacitor C_i. This provides the equivalent circuit shown in Fig. 24-13. Now use the scope to measure the generator output (V_{gen}) and then the amplifier input (V_i) as you adjust R_i so that $V_{gen} = 2V_i$. You may have to measure these voltages and adjust R_1 several times because the generator's output may change when R_1 is adjusted. When $V_{gen} = 2V_i$, the voltage across R_1 will equal the voltage across Z_i. Since R_1 and Z_i are in series and have equal voltage drops, their oppositions must also be equal. Finally, remove R_1 (without changing its setting) and measure that part of its resistance which was in the circuit. This is the value of Z_i.
k. Under what condition did the greatest error between calculated value and measured value occur? Why?
l. Compare Fig. 24-12(b) and (c). What happened to Z_i as A_V was increased by bypassing R_E?

2. CC amplifier
   a. Will the input signal and output signal in Fig. 24-14 be in-phase or 180° out-of-phase?
   b. Calculate and record Z_i and A_V for the circuit in Fig. 24-14. Notice that R_E and R_L are in parallel for the signal (ac) currents.
   c. Calculate and record the quantities listed in the table in Fig. 24-15(a).
   d. Construct the circuit shown in Fig. 24-14 noting especially the polarities of V_{CC} and the capacitors. Also note that a PNP transistor is used. Measure and record the quiescent voltages and current specified in Fig. 25-15(a).
   e. If there is more than a 15 percent disagreement between the values recorded in steps 2c and d, repeat both steps.
   f. Measure and record the Z_i of the CC amplifier. Why might this value be greater than the calculated value?
   g. Measure, and record in Fig. 24-15(b), the voltage gain of the CC amplifier under the three conditions specified in Fig. 24-15(b).
   h. How much change was there in A_V as the load condition was changed from no-load to 560 Ω?
   i. Compare Fig. 24-15(b) to Fig. 24-12(b). Would the unbypassed CE amplifier or the CC amplifier work best for driving loads that vary from 10 kΩ to 500 Ω? Why?
   j. Would the voltage gain of Fig. 24-14 still be approximately 1 if R_L were reduced to 51 Ω? Why?

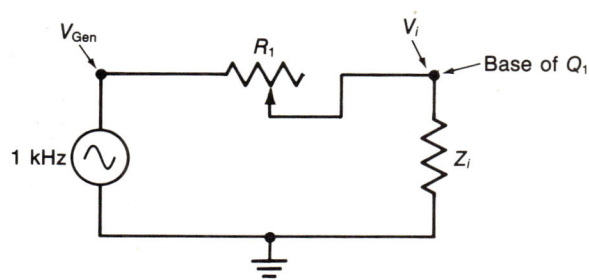

FIGURE 24-13 Measuring Z_i of an amplifier.

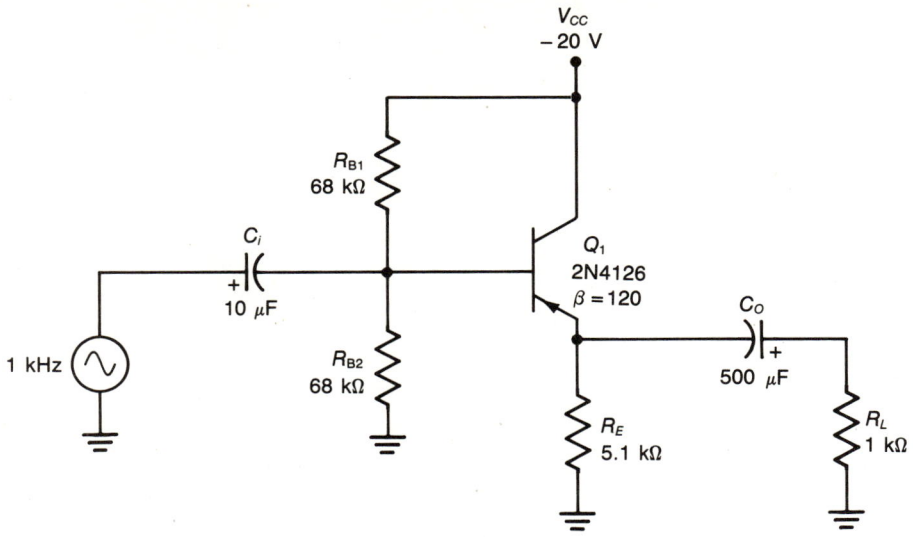

FIGURE 24-14 CC amplifier.

	Quiescent values	
	Calculated	Measured
V_B		
V_E		
V_{CE}		
I_C		

(a)

	Measured voltage gain
No-load	
$R_L = 1\,k\Omega$	
$R_L = 560\,\Omega$	

(b)

FIGURE 24-15 Data tables for the circuit in Fig. 24-14.

ACTIVITY 24-4 FET AMPLIFIERS

Introduction

In this activity you will construct and test common-source amplifiers using a JFET and a MOSFET. These circuits will highlight some of the major differences (biasing, input impedance, and voltage gain) between BJT and FET amplifiers. Also, in this activity, you will deal with amplitude and frequency distortion.

Supplies

- (1) Power supply, 20-V dc, floating output
- (1) JFET, 2N5458 or equivalent
- (1) MOSFET, 2N3797 or equivalent
- (1) Potentiometer, 1-MΩ, $\frac{1}{2}$-W, ±5%, linear
- (1) Resistor, 820-Ω, $\frac{1}{2}$-W, ±5%
- (1) Resistor, 1-kΩ, $\frac{1}{2}$-W, ±5%
- (1) Resistor, 2.2-kΩ, $\frac{1}{2}$-W, ±5%
- (1) Resistor, 5.1-kΩ, $\frac{1}{2}$-W, ±5%
- (1) Resistor, 10-kΩ, $\frac{1}{2}$-W, ±5%
- (1) Resistor, 180-kΩ, $\frac{1}{2}$-W, ±5%
- (1) Resistor, 910-kΩ, $\frac{1}{2}$-W, ±5%
- (1) Capacitor, 0.01-μF, 25 WVDC
- (1) Capacitor, 10-μF, 25 WVDC
- (1) Capacitor, 500-μF, 50 WVDC
- (1) Signal generator, audio range
- (1) Oscilloscope, ×10 probe, external trigger
- (1) Multimeter (DMM or VOM)

Procedure

1. JFET, common-source amplifier
   a. Construct the circuit shown in Fig. 24-16, but do not connect C_S or R_L at this time. Double check the polarity of C_o and V_{DD}. Turn on the V_{DD}

supply, but leave the signal generator turned off. Measure and record V_G, V_S, V_D, and V_{GS}.

b. Measure and record the value of I_D.

c. Observing the polarity of C_S, connect it and R_L into the circuit. Should this change any values measured in step 1(a)? Why? Check your first answer by measuring these voltages again.

d. Turn on the signal generator and apply an input signal. Using the oscilloscope, measure the input and output signal voltage; then, figure and record the voltage gain. How does this gain compare to the gain of the CE amplifier with a fully bypassed R_E (refer to Activity 24-3 if necessary).

e. If R_L is changed to 1 kΩ, what should happen to A_V? Change R_L to 1 kΩ; then measure and record the new A_V.

f. Using the 1-MΩ potentiometer and the technique described in Activity 24-3, measure and record Z_i for this circuit. How does this compare to the Z_i of the CE amplifier when R_E is fully bypassed? Should removal of C_S increase Z_i for this circuit? Check your answer, but reconnect C_S when you are through.

g. Now we will look at amplitude distortion. Turn off the signal generator. Set the oscilloscope to dc coupling and 0.2 V/div. (with the ×10 probe, this is equivalent to 2 V/div.). With the vertical position control, locate the trace at the bottom of the screen. Connect the vertical input probe to the drain and now adjust the vertical position control so that the trace is exactly in the center of the screen. This allows you to observe how much the negative and positive half-cycles are amplified when a signal is applied.

Turn on the signal generator and increase its output until the output signal fills the screen or until no evidence of signal clipping is present. Is one half-cycle amplified more than the other half-cycle? If yes, which half-cycle? Could this result have been predicted from the drain curves of a JFET and the 180° phase shift of a CS amplifier?

h. Leave everything exactly as in step 1g but remove C_S. What happened to the amplitude of the output signal? Why? Increase the input signal until the output signal again fills the screen or until no evidence of clipping is present. Has negative feedback reduced the amplitude distortion?

i. Return the scope to ac coupling. Remove R_L and connect C_S back into the circuit. Adjust the signal generator so that the signal voltage on the gate is exactly 0.1 V. Next, measure the output voltage and record the gain as "A_V at 1 kHz." Now, reduce the signal generator's frequency to 100 Hz and adjust the output voltage level until the signal on the gate is exactly 0.1 V. Again, measure the output voltage and record the gain as "A_V at 100 Hz." Has the gain increased or decreased? Does this change in gain represent frequency distortion? The results you have observed are due to an inadequate amount of input-coupling capacitance. This low-frequency distortion could be eliminated by increasing the value of C_i.

2. Depletion-type MOSFET amplifier

a. Refer to Fig. 24-17. What value of voltage would you predict for V_G, V_{GS}, and V_{RL}?

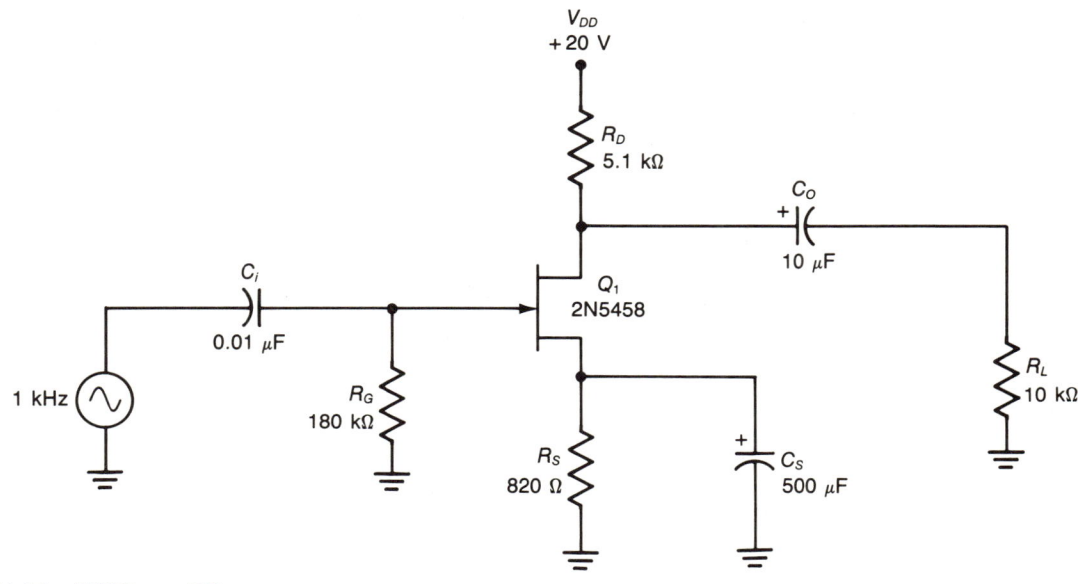

FIGURE 24-16 JFET amplifier.

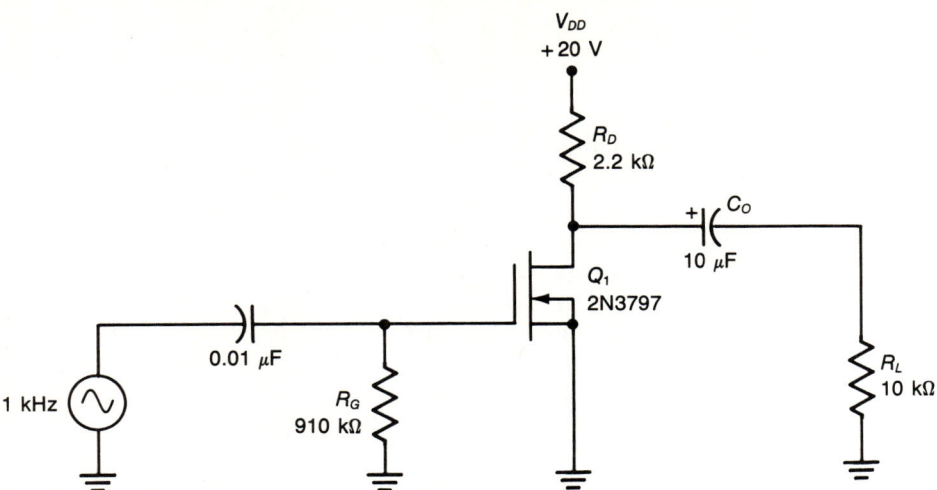

FIGURE 24-17 MOSFET amplifier.

b. Observing the procedures for handling a MOSFET (see Activity 24-2), construct the circuit shown in Fig. 24-17. Now, apply V_{DD} and check the voltages for which you predicted values in step 2a. Do they agree with your predictions?
c. What value of Z_i would you expect for Fig. 24-17? Measure Z_i and record its value.
d. Measure and record the gain of this amplifier at 1000 and at 100 Hz. Is there any evidence of frequency distortion? Why?

ACTIVITY 24-5   CE AMPLIFIER DESIGN

Introduction

Most of the formulas, assumptions, and procedures used in analyzing an amplifier can also be used in designing an amplifier. In this activity you are required to select most of the component values needed to produce an amplifier with specified characteristics.

Supplies

(1) DMM
(1) Oscilloscope
(1) Power supply, 0- to 25-V dc

Miscellaneous components selected from the Materials List in this manual.

Procedure

1. Determine the values of the unspecified components in Fig. 24-18 required to produce an amplifier with the following low-frequency and static characteristics:
   a. $I_C = 0.95$ mA
   b. $A_V = 6.9$
   c. Z_i between 4 kΩ and 6 kΩ
2. After constructing and testing your amplifier, write an activity report which meets or exceeds the requirements given in Appendix 1.

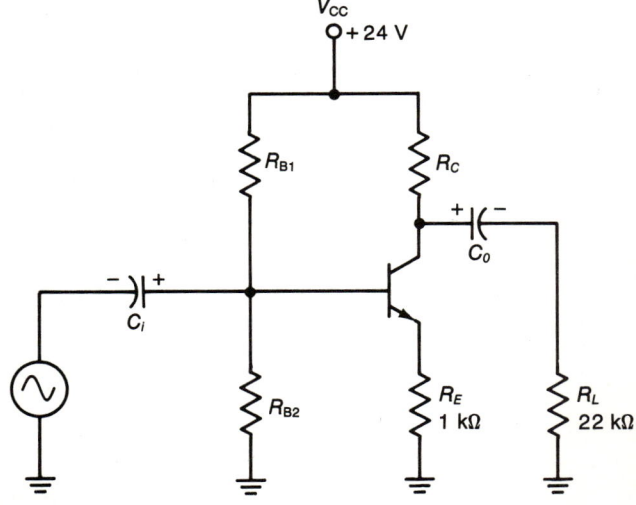

FIGURE 24-18   CE amplifier to be completed.

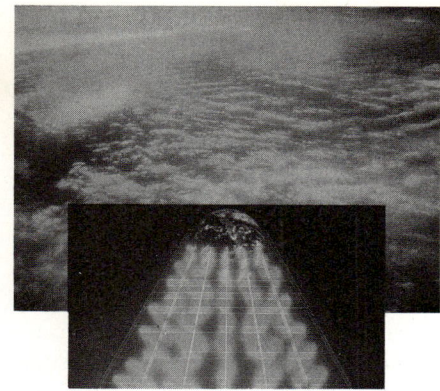

Chapter 25

Integrated Circuits and Operational Amplifiers

ACTIVITY 25-1 DIFFERENTIAL AMPLIFIER

Introduction

During this activity you will study the quiescent and signal characteristics of a differential amplifier. After constructing the amplifier circuit, you will measure a number of electrical quantities, for both quiescent and signal conditions, which should verify the results you obtained from your mathematical analysis of the circuit.

Supplies

- (2) Power supplies, 0- to 25-V dc, floating output
- (2) BJT's, 2N4124 or equivalent
- (2) Resistors, 150-kΩ, $\frac{1}{2}$-W, ±5%
- (1) Resistor, 120-kΩ, $\frac{1}{2}$-W, ±5%
- (1) Resistor, 68-kΩ, $\frac{1}{2}$-W, ±5%
- (1) Multimeter, 10-MΩ input R (DMM or electronic MM)
- (1) Ammeter, with microampere range (DMM or VOM)
- (1) Oscilloscope, ×10 probe, external trigger
- (1) Signal generator, audio range

Procedure

1. Quiescent conditions
   a. Is the input in Fig. 25-1(a) a differential input or a common-mode input?
   b. Assume G_1 has zero internal resistance and that it provides a dc path to ground. Calculate the quiescent values for the voltages and currents listed in Fig. 25-2(a). Record these values in the first column of the table in Fig. 25-2(a).
   c. Construct the circuit shown in Fig. 25-1(a). With the generator turned off, measure, and record in Fig. 25-2(a), the quiescent voltages and currents for the circuit. If the measured and calculated values disagree by more than 20 percent, recheck the circuit and the measurements. If there is still more than a 20 percent disagreement between V_{C1} and V_{C2}, the transistors are extremely mismatched. Try changing the transistors one at a time until the collector voltages are within 20 percent of each other.

2. Common-mode input
   a. Turn on the signal generator and set it for a 1 kHz, 1 V$_{p-p}$ output. This provides a 1 V$_{p-p}$ common-mode input signal. Use the oscilloscope (×10 probe) and measure the single-ended signal outputs at outputs 1 and 2. (Measure between the output terminal and ground.) Record the measured output voltages in Fig. 25-2(b). If $R_{C1} = R_{C2}$ and Q_1 is identical to Q_2, should these outputs be equal?
   b. Should the output at 1 be in-phase or out-of-phase with the output at 2 in Fig. 25-1(a)? Check your answer with the oscilloscope using the external trigger function.
   c. Using the DMM on the ac V function or the VOM on the ac output V function, measure the signal voltage between the collector of Q_1 and the collector of Q_2. Record the value measured in Fig. 25-2(b) in the "dual-ended output" row. Is the dual-ended output greater than or less than the single-ended output when the input is a common-mode signal?
   d. Using the larger of the two single-ended outputs and the 1 V$_{p-p}$ input, determine A_{CM} and record its value in the measured column of Fig. 25-2(b). Use the general gain formula which is $A_V = V_o/V_i$. Is A_{CM} less than one? If not, check the measurements and the circuit?

131

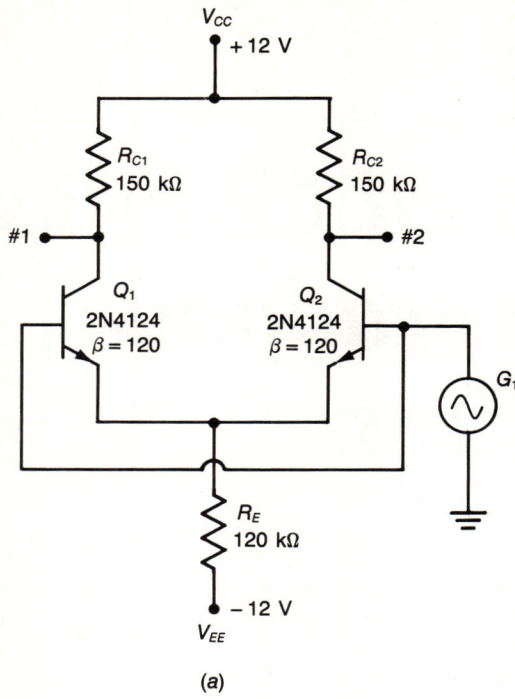

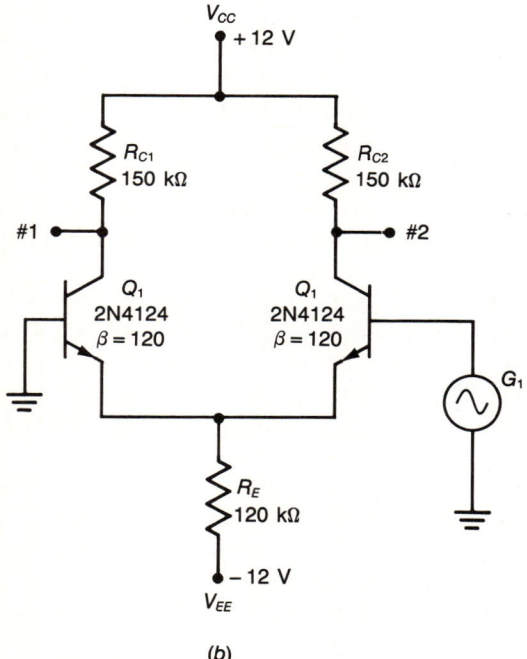

FIGURE 25-1 DA circuits.

Adjust G_1 so that it provides an input signal of 0.02 $V_{p\text{-}p}$ (measured with the ×1 probe on the oscilloscope) at 1 kHz. Make the measurements needed to complete the "measured" column in Fig. 25-2(c). (Use the ×10 probe when measuring output signal voltages.) In determining the single-ended output gain (A_D), use the larger of output 1 or output 2. In determining the Z_i, insert a 68-kΩ resistor between the base of Q_2 and the output of G_1. Then increase the output of G_1 as much as possible without distorting the output from terminal 1 or terminal 2. Now measure the voltage out of G_1 (V_{G_1}) and the voltage between the base of Q_2 and ground (V_{BQ2}). Finally,

	Quiescent values	
	Calculated	Measured
V_{C1}		
V_{C2}		
V_{B1}		
V_{B2}		
I_{RE}		
I_{C1}		
I_{C2}		

(a)

	Common mode (1.0 $V_{p\text{-}p}$ input signal)	
	Calculated	Measured
Output #1		
Output #2		
Dual-ended output		
A_{CM}		

(b)

	Differential mode (0.02 $V_{p\text{-}p}$ input signal)	
	Calculated	Measured
Output #1		
Output #2		
A_D		
Z_i		

(c)

FIGURE 25-2 Data tables for the circuits in Fig. 25-1.

e. Using only the values given on the diagram in Fig. 25-1(a), calculate and record the voltages and gain needed to complete the first column in Fig. 25-2(b).

3. Differential input
   a. Change the input connection to the DA so that it matches the diagram in Fig. 25-1(b). This circuit provides a single-ended differential input.

you can determine Z_i from your measured values. Since $V_{68\,k\Omega} = V_{G_1} - V_{BQ2}$, the appropriate formula is

$$Z_i = \frac{V_{BQ2} \times 68\text{ k}\Omega}{V_{G_1} - V_{BQ2}}$$

b. Should the output at 1 or the output at 2 be out-of-phase with the input? Using the external trigger function of the oscilloscope, check your answer. Is output 1 in-phase or out-of-phase with output 2?

c. Calculate all the quantities called for in the "calculated" column of Fig. 25-2(c). In making these calculations, use only the information provided on the diagram in Fig. 25-1(b).

d. Using measured values from Fig. 25-2(b) and (c), determine the CMRR for this differential amplifier. Record this value of CMRR. Using calculated values from these same two figures, calculate and record the CMRR. In addition to measurement error and component tolerances, why might these two values of CMRR be quite different?

e. Why might your calculated value of Z_i be considerably smaller than your measured value of Z_i?

ACTIVITY 25-2 OPERATIONAL AMPLIFIER

Introduction

In this activity you will analyze the operation of two popular op-amp circuits—the inverting and the non-inverting amplifier. The circuits will be tested for their response to both dc and ac signals. Such characteristics as bandwidth, gain, input impedance, and slew rate will be tested.

Supplies

(2) Power supplies, 0- to 25-V dc, floating output
(1) Op-amp, 741, 8-pin, DIP
(1) Potentiometer, 1-kΩ, $\frac{1}{2}$-W, $\pm 5\%$, linear
(1) Resistor, 1-kΩ, $\frac{1}{2}$-W, $\pm 5\%$
(1) Resistor, 5.1-kΩ, $\frac{1}{2}$-W, $\pm 5\%$
(2) Resistors, 10-kΩ, $\frac{1}{2}$-W, $\pm 5\%$
(1) Resistor, 68-kΩ, $\frac{1}{2}$-W, $\pm 5\%$
(1) Resistor, 120-kΩ, $\frac{1}{2}$-W, $\pm 5\%$
(1) Resistor, 910-kΩ, $\frac{1}{4}$-W, $\pm 5\%$
(1) Multimeter, DMM or VOM
(1) Oscilloscope, $\times 10$ probe, external trigger
(1) Signal generator, audio range

Procedure

1. Inverting amplifier
   a. Refer to Fig. 25-4(a). The inverting amplifier is shown inside the dotted lines. R_3 and R_4 form an adjustable voltage divider so that a dc voltage can be applied to the inverting amplifier input. When R_1 is adjusted for zero input, what values of dc voltage would you expect at pins 2 and 6? Record these expected values in Fig. 25-5(a). Build the circuit in Fig. 25-4(a), but omit both jumper wires. (See Fig. 25-3 for pin connections on the 741.) Adjust R_4 toward its grounded end. Apply power; measure the dc voltage at pins 2 and 6; record these measured voltages in Fig. 25-5(a). Are these voltages less than 100 mV? If not, check your circuit!
   b. Connect a jumper from $+V_{CC}$ to R_3. Adjust R_4 to apply +1 V to the input of the inverting amplifier. Complete the calculated column of Fig. 25-5(b), and then make the measurements, with a dc voltmeter, needed to complete the measured column of Fig. 25-5(b). Remember that the "$-$ input V" is measured between pin 2 and ground. Do the measurements you made indicate that pin 2 is at "virtual ground?" Within the tolerances of R_1 and R_F, does the value of A_V

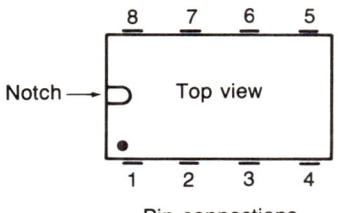

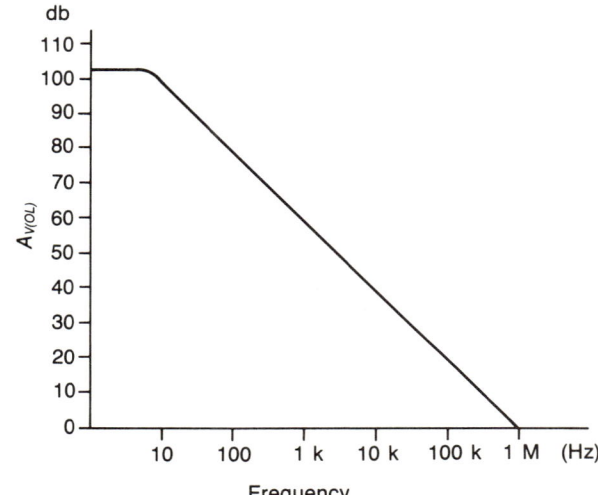

FIGURE 25-3 Data for a 741 op amp.

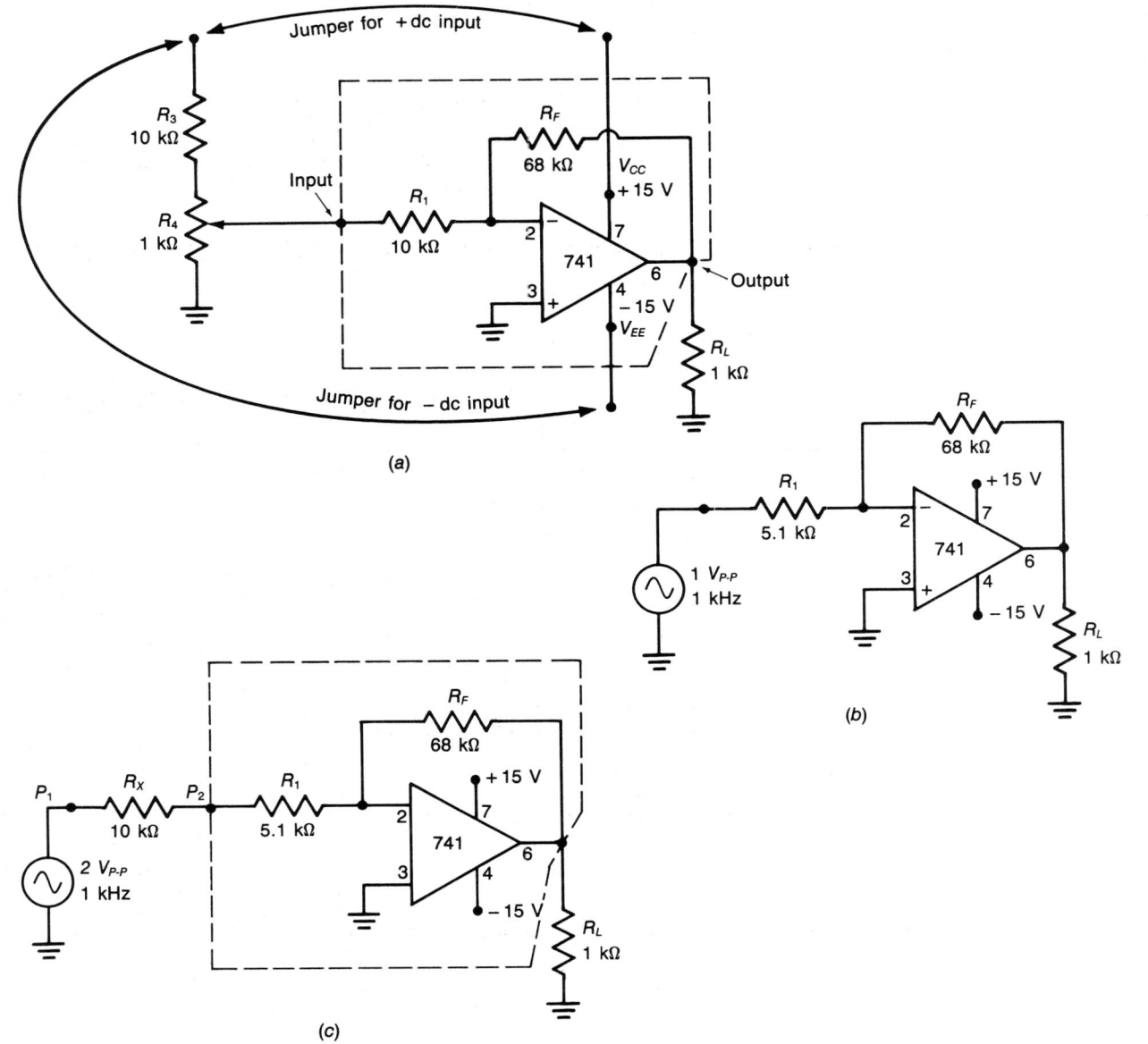

FIGURE 25-4 Inverting op-amp circuits.

obtained from the measurements indicate that $A_V = R_F/R_1$?

c. Remove the jumper from $+V_{CC}$ to R_3 and install a jumper from $-V_{CC}$ to R_3. Also, change R_1 to 5.1 kΩ. Now adjust R_4 so that the input to the amplifier is -0.5 V. What value of voltage would you expect at pin 6 under these conditions? Complete both columns of Fig. 25-5(c).

d. What is the phase relationship between the input and output of an inverting amplifier?

e. For the circuit diagram of Fig. 25-4(b), complete the calculated column in Fig. 25-5(d). Use the frequency response graph in Fig. 25-3 to determine $A_{V,OL}$ at 1 kHz. Remember that even though pin 2 is said to be at virtual ground, there is a small (millivolt) signal on pin 2. It is calculated by the formula

$$-\text{input } V = \frac{\text{output } V}{A_{V,OL}}$$

Construct the circuit shown in Fig. 25-4(b) and make the measurements needed to complete the measured column of Fig. 25-5(d). You may want to use the ×1 probe to measure the $-$input V. Measurements needed to determine Z_i can be obtained by adding R_x as shown in Fig. 25-4(c). The theory behind this indirect method of measuring Z_i has been discussed in previous activities. After measuring the voltages at V_{P1} and V_{P2} in Fig. 25-4(c), use the formula

$$Z_i = \frac{V_{P2} R_x}{(V_{P1} - V_{P2})}$$

f. What determines the Z_i of an inverting amplifier?

Quiescent values $R_1 = 10\ k\Omega,\ R_F = 68\ k\Omega$	
Calculated	Measured
−Input V	
Output V	

(a) No input

With +1 V input $R_1 = 10\ k\Omega,\ R_F = 68\ k\Omega$	
Calculated	Measured
−Input V	
Output V	
A_V	

(b) +dc input

With −0.5 V input $R_1 = 5.1\ k\Omega,\ R_F = 68\ k\Omega$	
Calculated	Measured
−Input V	
Output V	
A_V	

(c) −dc input

Input = 1 $V_{p\text{-}p}$ at 1 kHz $R_1 = 5.1\ k\Omega,\ R_F = 68\ k\Omega$	
Calculated	Measured
−Input V	
Output V	
A_V	
$A_{V,OL}$	
Z_i at 1 kHz	

(d) ac input

FIGURE 25-5 Data tables for the circuits in Fig. 25-4.

g. Would the A_V of the circuit in Fig. 25-4(b) be as large at 10 kHz as it was at 1 kHz? Check your answer by measuring A_V at 10 kHz.

h. Still using the circuit of Fig. 25-4(b) with a 1 $V_{p\text{-}p}$ input signal, increase the frequency of the input signal until the output just turns into a complete triangular waveform. This distortion is slew-rate distortion. Be sure that both the X axis and the Y axis of the oscilloscope are calibrated. Now, using the peak-to-peak value of the waveform and the time between the negative and positive peaks, calculate the slew rate of the 741 op amp. Record this value. The typical slew rate for a 741 op amp is 0.5 V/μs. Is the value you determined within 25 percent of this value?

i. If the amplitude of the input signal is reduced, should the slew rate distortion disappear? Does it?

2. Noninverting amplifier
   a. Should the A_V for the circuit in Fig. 25-6(a) be greater than or less than the A_V for the circuit of Fig. 25-4(b)?
   b. Compare the impedances of the two circuits specified in step 2a.
   c. Is pin 2 in Fig. 25-6(a) at virtual ground?
   d. Using the circuit and values specified in Fig. 25-6(a) and the frequency graph in Fig. 25-3, complete the calculated column in Fig. 25-7.
   e. Construct the circuit shown in Fig. 25-6(a) and make the measurements needed to complete the measured column in Fig. 25-7. A couple of reminders: (1) use external trigger to measure phase shift; and (2) the edge of the bandwidth is the frequency at which the gain has dropped 3 db below the gain at 1 kHz. When the output voltage drops to 70.7 percent of its former value, a 3-db drop has occurred.
   f. Because of the limited input impedance of most oscilloscopes and the very high input impedance of the noninverting amplifier, you will not be able to accurately measure the Z_i of the amplifier in Fig. 25-6. However, to illustrate the high Z_i of this circuit *and* the loading effect of an oscilloscope in a high-impedance circuit, connect the circuit of Fig. 25-6(b) and attempt to measure Z_i. Record the value of Z_i obtained by the measured voltages at $P1$ and $P2$.

   Figure 25-6(c) shows an equivalent circuit which explains why the value of Z_i obtained from Fig. 25-6(b) was very low compared to its actual value. In this equivalent circuit, C_W is the wiring capacitance of the circuit. Wiring capacitance refers to the capacitance between conductors, components, etc. Its exact value depends on the wiring techniques used for the circuit. Although C_W may be only 1 pF, its reactance at 10 kHz is very significant when it is shunting (in parallel with) megohms of input impedance. C_O and R_O in Fig. 25-6(c) represent the input capacitance and input resistance, respectively, of a typical oscilloscope with an ×10 probe. From Fig. 25-6(c), you can see that how the signal voltage division in the circuit is controlled by R_X and C_O rather than by R_X and Z_i.
   g. Why did the indirect method of measuring Z_i give reasonably accurate results for the inverting amplifier?

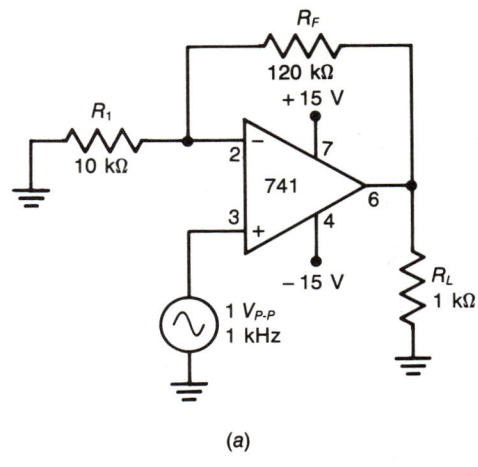
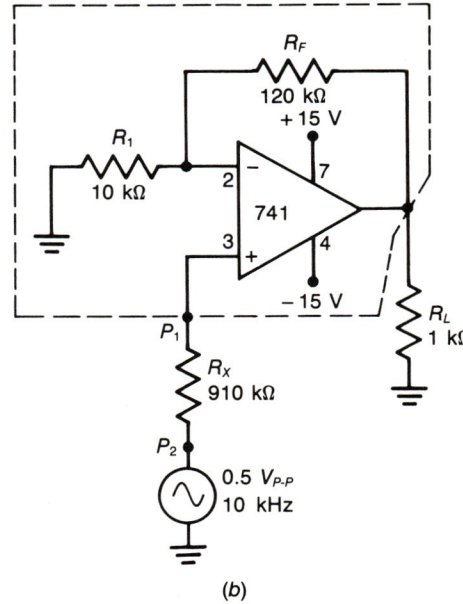
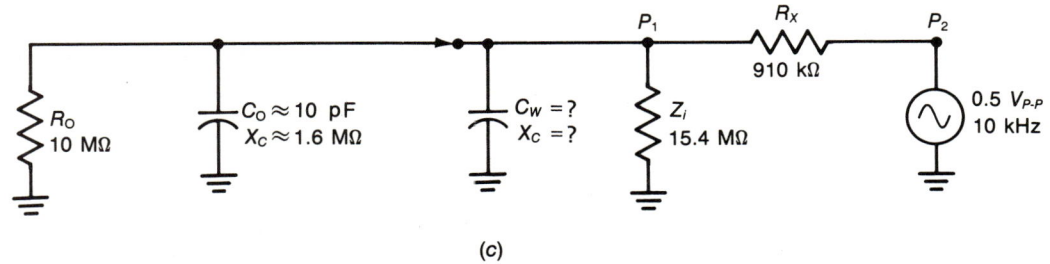

FIGURE 25-6 Noninverting op-amp circuits.

	Input = 1 V_{p-p} at 1 kHz R_1 = 10 kΩ, R_F = 120 kΩ	
	Calculated	Measured
+Input V		
−Input V		
Output V		
A_V		
Phase shift		
Bandwidth		
Z_i at 10 kHz		

FIGURE 25-7 Data table for the circuits in Fig. 25-6.

ACTIVITY 25-3 OP AMP DESIGN

Introduction

This activity requires you to design two circuits using the 741 op amp. The pin connections and the response curve for the 741 op amp are provided in the previous activity (Fig. 25-3).

Supplies

(2) Power supplies, 0- to 25-V dc
(3) D cells or 2 more dc power supplies
(1) Oscilloscope
(1) Signal generator, audio range
(1) Op amp, 741

Miscellaneous resistors selected from the Materials List in this manual.

Procedure

1. Design, construct, and test:
   a. An op amp circuit with a bandwidth of 37.3 kHz and an input impedance of 820 Ω.
   b. A two-input summing amplifier that provides a V_o of -3.3 V when input 1 = -1.5 V, input 2 = $+3$ V, $R_F = 2.2$ kΩ, and $R_1 = 3$ kΩ.
2. Submit an activity report following the guidelines in Appendix 1.

Chapter 26

Introduction to Digital Circuits

ACTIVITY 26-1   DECISION-MAKING GATES

Introduction

In this activity you will use IC logic gates to verify the truth tables for several types of logic gates. In the process you will gain experience in handling and testing TTL ICs. You will learn how to use NAND or NOR gates to produce AND, OR, and NOT functions. You will use an LED to determine if a logic state is 1 or 0.

Supplies

(1) Power supply, 5-V dc
(1) Digital IC, 7400
(1) Digital IC, 7402
(3) Resistors, 220 Ω, ¼-W, ±5%
(2) Switches, SPDT (any current and voltage rating)
(3) LED's, T-1¾, 40-mA maximum

Procedure

1. Digital IC familiarization

   Figure 26-1(a) shows how to identify the pin numbers on a DIP IC. On some ICs, only a U-shaped notch is depressed into the plastic package on the end of the package where pin 1 and the last pin (pin 8, 14, 16, 24, or 40) are located. Pin 1 is located on that corner of the notched end which allows all pins to be sequentially counted in a counterclockwise direction without crossing the notched end. On some ICs a depressed dot is located next to pin 1.

   Figure 26-1(b) shows the pin assignments for a 7400 IC which houses 4 two-input NAND gates. The GND (ground) and V_{CC} (collector supply voltage) are common to all four NAND gates. GND goes to the negative terminal and V_{CC} goes to the positive terminal of a 5-V supply.

   The pin assignments for an IC (7402) with 4 two-input NOR gates are shown in Fig. 26-1(c). Note that the ground and V_{CC} pins on the chip (IC) are pins 7 and 14, respectively. Many 14-pin DIP ICs use these two pins for ground and V_{CC}.

   It is very easy to bend and/or break the pins on an IC package—especially when removing the IC

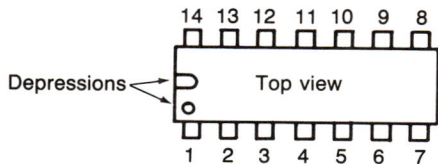

(a) Count pins in a counter-clockwise direction

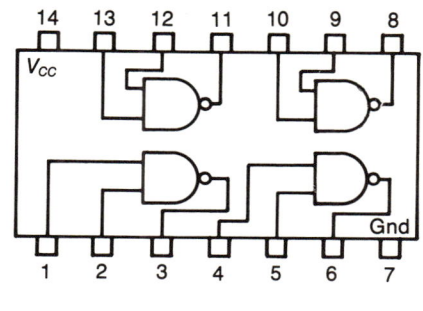

(b) 7400

(c) 7402

FIGURE 26-1 IC pin connections.

from a socket. Use the procedure demonstrated by your instructor.

Digital ICs are easily destroyed by excess V_{CC} voltage. Be certain your power supply is providing no more than 5.5 V dc and is well-filtered.

2. NAND gates
   a. The circuit in Fig. 26-2(a) can be used to test the operation of a NAND gate. The series resistor and LED connected to each switch and input lead indicates whether the input is logic 1 or logic 0. The LED will glow when the input lead is high (logic 1). D_3 indicates the logic level of the output. Since D_3 returns to V_{CC}, it will glow when the output is logic 0.

   Construct the circuit in Fig. 26-2(a). Be sure to apply 5 V between pin 7 and pin 14 to power the IC. Switch S_1 and S_2 so that both inputs (A and B) are low. Is D_3 glowing? Is the output (X) a logic 1 or a logic 0?

   b. If S_1 is changed so that input A is high, will output X be high or low? Check your answer by changing S_1 and observing D_3.
   c. Under what input conditions will D_3 glow? Will the output be high or low at this time? Check your answer by changing S_1 and/or S_2 to provide these conditions.
   d. Change S_1 and/or S_2 to provide any possible input combination of 1's and 0's that you have not already used and observe the effect on the output. Use the data you have collected or observed to construct a truth table for a NAND. Label the inputs A and B and output X. List the input combinations in ascending order.
   e. Disconnect S_1, R_1, and D_1 from input A (pin 1). Now, while observing the output, flip S_2 back and forth so that input B is alternately 1 and 0. Does the open-circuited input A act like it has a logic 1 or a logic 0 applied? Although open leads on TTL ICs behave in the way just demonstrated, they should be connected to a logic 1 when not in use. This greatly reduces the probability of an induced noise spike driving the input to a temporary logic 0.

3. NOR gate
   a. Construct the circuit shown in Fig. 26-2(b). Set the switches so that both inputs are logic 1. Is the output low or high?
   b. Does a NOR (or a NAND) have an active low or an active high output?
   c. Under what input conditions will the output of this circuit be active?
   d. Construct and record a truth table for this circuit. Verify your truth table by using S_1 and S_2 to input the various combinations indicated in the truth table. Remember that D_3 glows when the output is low, while D_1 and D_2 glow when their respective inputs are high.

4. Universal property of NAND and NOR gates
   a. NANDs and NORs are sometimes referred to as universal gates or gates with universal properties because either type can be used for the NOT, AND, and OR functions. Figure 26-3 shows how a NAND, or NANDs, can be connected for each of these three functions. The

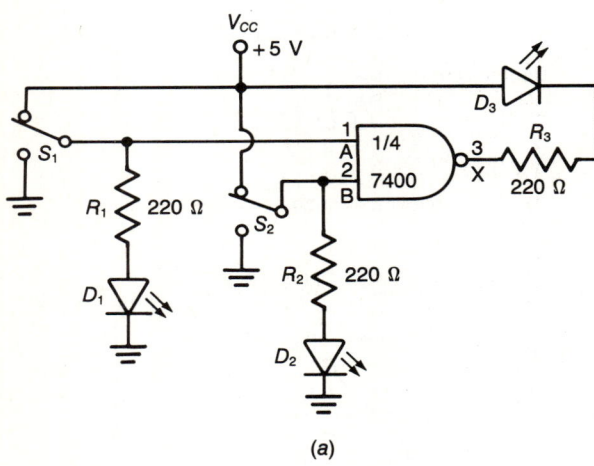

(a)

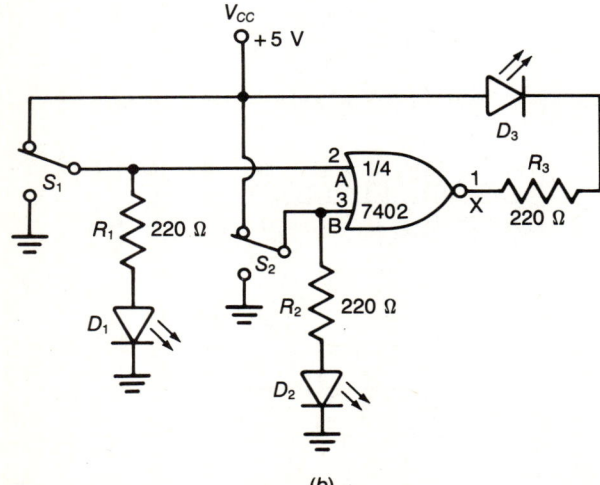

(b)

FIGURE 26-2 NAND and NOR test circuits.

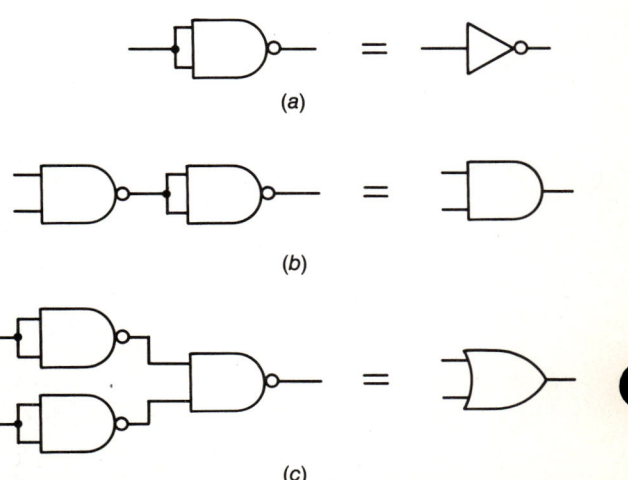

FIGURE 26-3 Universal property of NAND gates.

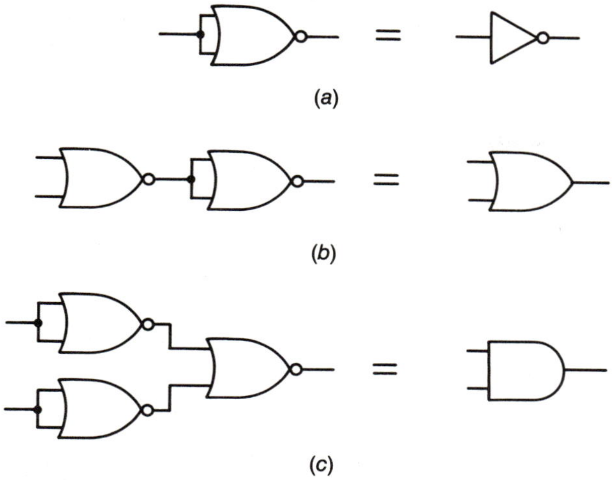

FIGURE 26-4 Universal property of NOR gates.

equivalent circuits using NOR gates are shown in Fig. 26-4.

Using the universal property of these gates can sometimes reduce the number of ICs needed for a logic design. For example, the boolean expression $X = \overline{\overline{AB}}$ can be implemented with two NAND gates contained in a single IC instead of a NAND gate (one IC) and a NOT gate (another IC).

b. Construct the circuit in Fig. 26-5. Notice that D_3 is returned to ground rather than V_{CC}. Thus, it will glow when the output is high (logic 1). While observing D_3, manipulate S_1 and S_2 to provide all possible input combinations. Construct and record the truth table for the circuit in Fig. 26-5.
c. Does the circuit in Fig. 26-5 perform the OR function?
d. Reconstruct the circuit in Fig. 26-5 using NOR gates instead of NAND gates. Construct and record the truth table for this new circuit. Now verify that the table is correct by changing the inputs with S_1 and S_2.
5. Review questions
   a. For a two-input NAND gate, how many input combinations produce a high output?
   b. How many NOR gates does it take to produce the AND function?
   c. How many NAND gates does it take to produce the AND function?
   d. What is the purpose of R_1, R_2, and R_3 in Fig. 26-5?

ACTIVITY 26-2 FLIP-FLOP CIRCUITS

Introduction

A flip-flop is a very versatile circuit. In addition to showing the characteristics of several types of flip-flops, this activity will demonstrate the use of flip-flops as counters and frequency dividers. Also included in this activity is the important concept of switch debouncing.

Supplies

(1) Power supply, 5-V dc
(1) Digital IC, 7400
(1) Digital IC, 7476
(1) Digital IC, 555
(3) Switches, SPDT (any current and voltage rating)
(3) Resistors, 220-Ω, $\frac{1}{4}$-W, ±5%
(2) Resistors 1-kΩ, $\frac{1}{4}$-W, ±5%
(1) Resistor, 2.2-kΩ, $\frac{1}{4}$-W, ±5%
(1) Resistor, 22-kΩ, $\frac{1}{4}$-W, ±5%

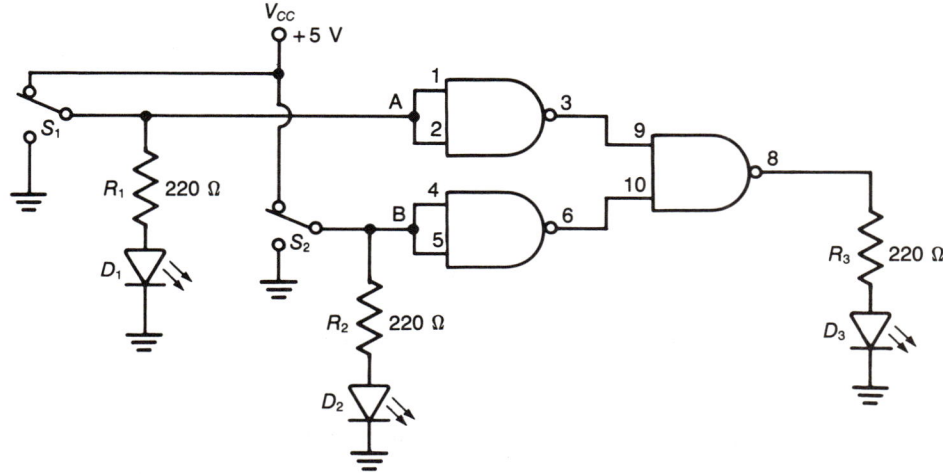

FIGURE 26-5 OR function using NAND gates.

(3) LED's, T 1¾, 40-mA maximum
(1) Capacitor, 0.1-μF, 25-V, $\pm 10\%$
(1) Capacitor, 0.01-μF, 25-V, $\pm 10\%$
(1) Oscilloscope, dc-coupled, external triggering

Procedure

1. Switch debouncing with $\overline{S}\,\overline{R}$ flip-flop
   a. When a switch, such as S_1 in Fig. 26-6, is closed, its contacts do not come together once and stay together. Instead, the contacts hit together, bounce open, and hit together again. The switch contacts may bounce a number of times before they settle down and remain closed. If one tries to count the number of switch closures with a digital counter, the counter will count each bounce as if it were a separate closure of the switch. Since the number of bounces varies from closure to closure, there is no way to know from the counter output how many times the switch has been opened and closed. Thus, the need for debouncing a switch.

   The resistors in Fig. 26-6 are called "pull-up" resistors because they pull up to a logic 1 the line to which they are connected unless that line is pulled down to a logic 0 by the grounded pole of S_1. Since S_1 is in the up position, $\overline{S}$ will be 0 and $\overline{R}$ will be 1. This condition sets the FF ($Q = 1$, $\overline{Q} = 0$). When S_1 is moved to the down position, there will be an instant when $\overline{S} = 1$ and $\overline{R} = 1$; but, nothing will happen at the output because 1,1 is a no-change condition for $\overline{S}\,\overline{R}$ flip-flop. The first instant the pole contacts the lower contact, $\overline{R}$ will be 0 and $\overline{S}$ will be 1. This will immediately reset the FF ($Q = 0$ and $\overline{Q} = 1$). The switch can now bounce many times but the output will not change because bouncing will merely drive the inputs to 1,1 which is a no-change condition.
   b. Construct the circuit shown in Fig. 26-6. Do not forget to connect the 5-V supply between pins 7 and 14. Connect the oscilloscope's vertical input to pin 1. Set the oscilloscope controls as follows: dc-coupled, 1 V/div. and 2 ms/div. Adjust the sweep stability for free running so that a trace appears on the oscilloscope's screen. Now operate S_1 four or five times while observing the oscilloscope screen. Next, move the oscilloscope's vertical input to pin 3 and again operate S_1. Is there evidence that the $\overline{S}\overline{R}$ FF has debounced the switch? If so, describe the evidence.

   Save this debouncer circuit. It will be used in the next two sections of this activity to provide clock pulses to the clocked flip-flops.

2. J-K flip-flop characteristics
   a. Construct the circuit shown in Fig. 26-7. The rectangle in Fig. 26-7 that is labeled "debounced switch" represents the circuit in Fig. 26-6. When power is applied to a clocked flip-flop, the flip-flop is randomly either set or reset. If the flip-flop in Fig. 26-7 is set, D_1 will glow.

   When the Q output of a flip-flop is high (logic 1), it may not be able to provide enough current to operate an LED and still remain at a logic 1. Thus, the NAND gate, connected to provide the NOT function, is used to drive the LED. The low output of the NAND gate can easily handle the current required to light the LED. Change S_2 and S_3 as necessary to make $J = 1$ and $K = 0$. Now operate S_1 one time; i.e., move the switch control to a new position and back to its original position. Is the flip-flop in the set or reset condition? Is D_3 glowing?
   b. Operate S_1 one more time. Did the output of the flip-flop change?
   c. Change J to 0 and K to 1. Did the output of the flip-flop change? Will the output change if S_1 is operated once? Check your answer by operating S_1 once.
   d. Will the output change if you operate S_1 again? Check your answer by operating S_1.
   e. Will the output of the flip-flop change when J and K are changed to $J = 1$ and $K = 1$ or to $J = 0$ and $K = 1$? Verify your answer by changing S_2 and S_3 as required.
   f. With $J = 0$ and $K = 0$, operate S_1 three or four times. Describe what happened to the output of the flip-flop.
   g. Repeat step f with $J = 1$ and $K = 1$.
   h. Disconnect pin 2 of the 7476 from V_{CC} and connect it to ground. Is the flip-flop now set or reset? Try to change the output of the flip-flop by using various J-K combinations and clocking the flip-flop with S_1. Were you able to change the output? Change pin 2 back to V_{CC}. Did this change the output?
   i. Describe what will happen when pin 3 is changed from V_{CC} to ground and S_2, S_3, and S_1 are manipulated. Check your answer by changing pin 3 to ground and operating the three switches. Return pin 3 to V_{CC}.
   j. List three ways to change a JK flip-flop from the reset to the set condition.

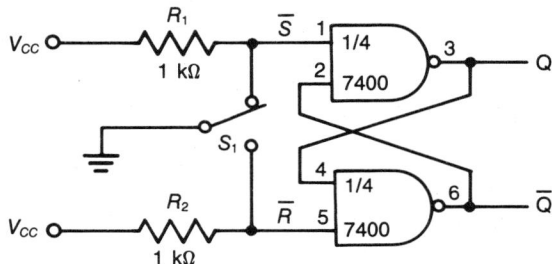

FIGURE 26-6 Switch debouncer.

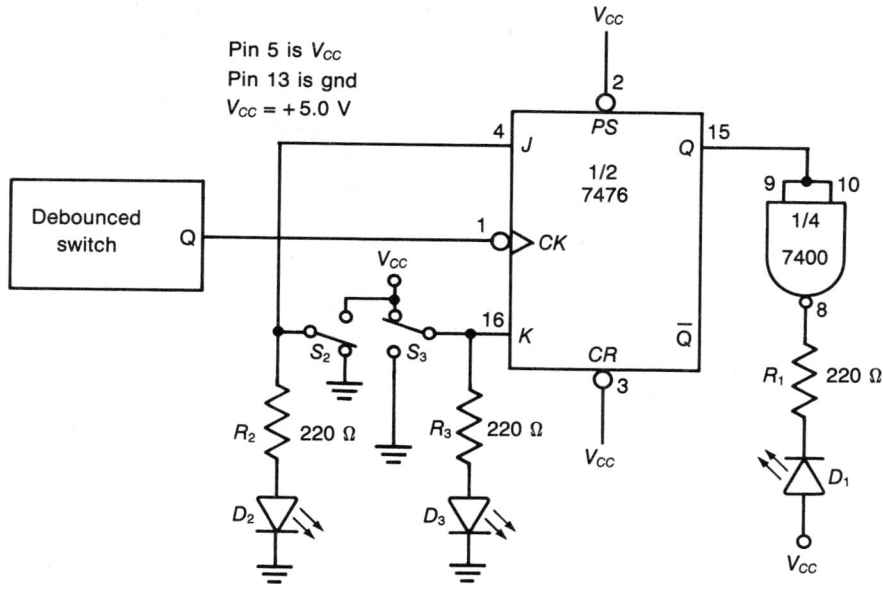

FIGURE 26-7 Circuit for testing a *J-K* flip-flop.

3. Counter operation
   a. The circuit in Fig. 26-8 uses two *JK* flip-flops to provide a modulus-4 counter. Each flip-flop is operated in the toggle mode by holding *J* and *K* at a logic 1. The counter will count the number of switch closures made by the debounced switch of Fig. 26-6. Construct the circuit shown in Fig. 26-8. Preset the counter by temporarily connecting both pin 2 and pin 7 to ground; reconnect both pins to V_{CC}. Both D_1 and D_2 should be glowing. This indicates that the counter is full (at its maximum count). When S_1 (the debounced switch) is operated once, what should happen to D_1 and D_2? Operate S_1 once to check your answer. What binary number is now in the counter?
   b. List the sequence of binary numbers the counter will display after each of the next four clock pulses (operations of S_1). Remember, the least significant bit (LSB) appears at the Q_1 (pin 15) output. Clock the counter four times to check your answer. The counter should now contain a binary 00.
   c. Suppose pins 3 and 7 are temporarily connected to ground and then returned to V_{CC}. What binary number would be in the counter? What is the base 10 (decimal) equivalent of this number?
   d. Next, change the circuit wiring so that pin 1 of the 7476 connects to pin 1 (instead of pin 3) of the 7400. S_1 will now clock the counter without being debounced. Operate S_1 six times. Did the counter follow its normal count sequence? Why? *Change the wiring back so that the switch is again debounced.*

4. Decoding a counter
   a. Figure 26-9 is the same as Fig. 26-8 except that

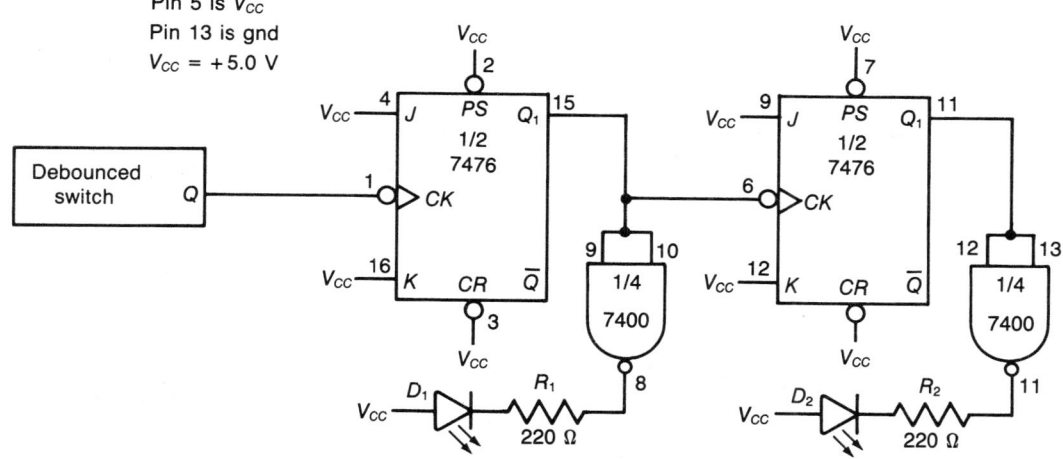

FIGURE 26-8 Modulus-4 counter.

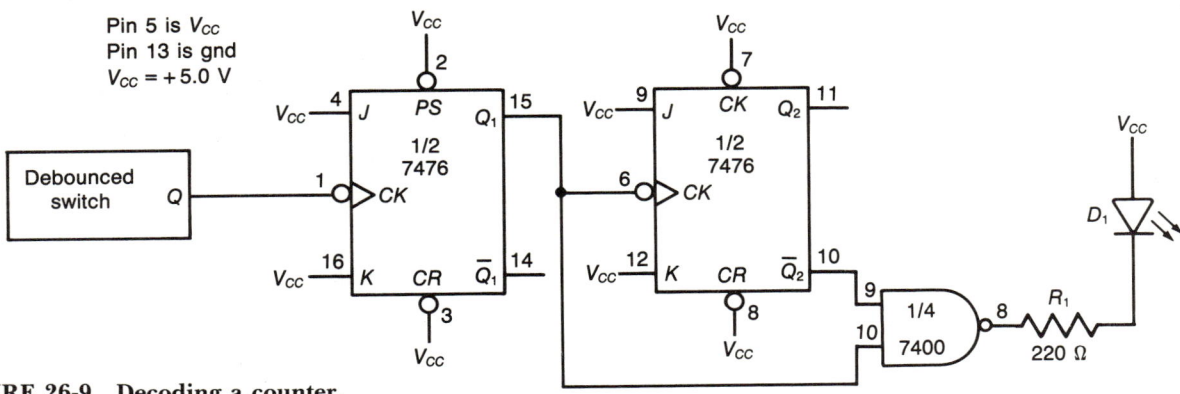

FIGURE 26-9 Decoding a counter.

(1) the LED circuits to indicate the logic state of Q_1 and Q_2 have been removed, and (2) a single-digit decoder has been added. (Since the logic state indicators have been removed, you will need to mentally keep track of the count.) The NAND gate connected to the outputs of the two flip-flops will decode one of the four binary numbers this counter can count. What is the decimal equivalent of the binary number which will be decoded (cause D_1 to glow)? Check your answer by first clearing the counter (pins 3 and 8 connected to ground; then back to V_{CC}), and clocking the counter with S_1.

b. If you wanted to decode 3 (11 binary), to which pins of the 7476 would you connect the NAND input leads? Check your answer by making the required wiring changes.

c. If you wanted to simultaneously decode all four numbers for this counter, how many two-input NAND gates would be required? How many 7400 ICs would be required?

5. Frequency division with a counter

a. The circuit in Fig. 26-10(a) uses the same counter as Fig. 26-9; however, the debounced switch has been replaced with a 555 IC. The 555 IC is connected as an astable (free-running)

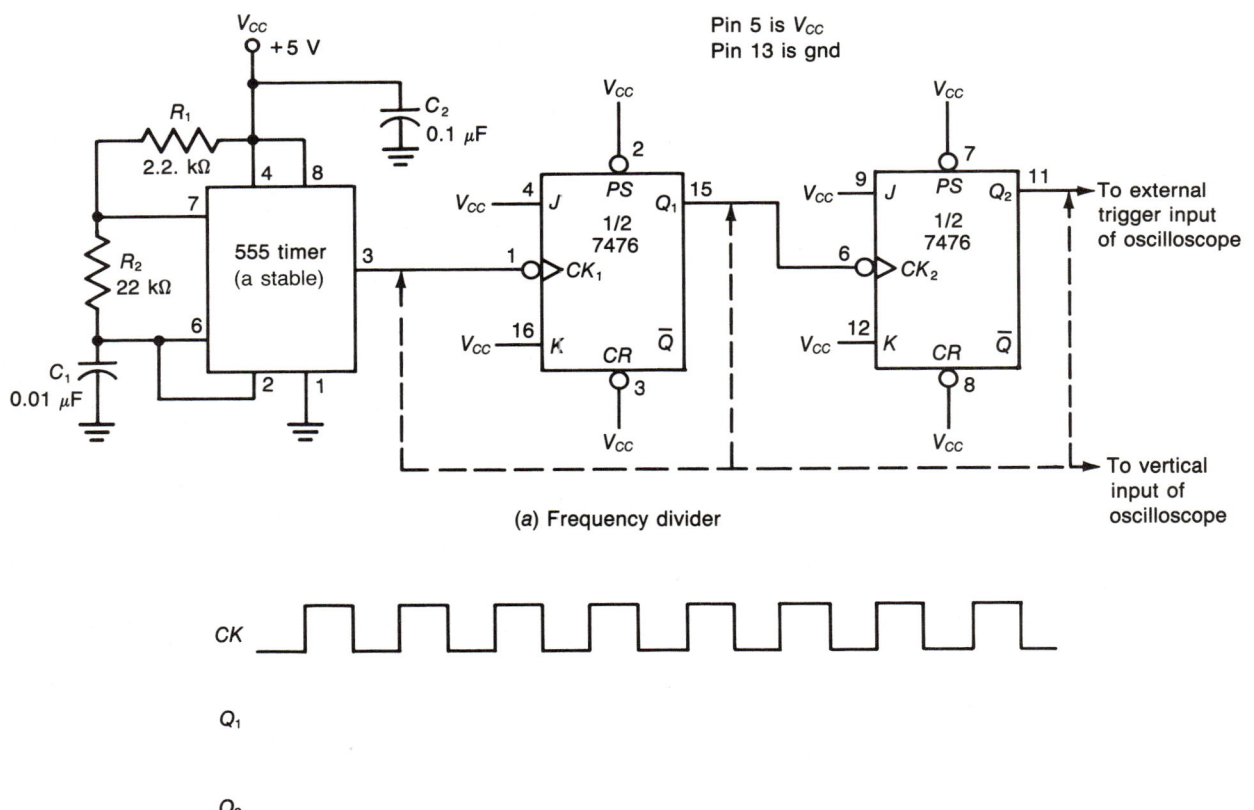

FIGURE 26-10 Frequency divider circuit.

multivibrator. An astable multivibrator generates a square-wave output. In this circuit, the square-wave output is the clock signal which drives the counter. The frequency and symmetry of the output of the 555 timer circuit are determined by the values of R_1, R_2, and C_1.

Construct the circuit in Fig. 26-10(a) and connect the vertical input of the oscilloscope to pin 1 of the 7476. Use dc coupling. Set the oscilloscope for negative-slope external triggering. Next, measure and record the period of the waveform. (Do not forget to have the sweep controls in their calibrated positions.) Now calculate and record the frequency of the clock signal. Finally, adjust the sweep control so that 8 cycles of the clock signal are displayed. Do not change the sweep controls again until instructed to do so.

b. Move the vertical input of the oscilloscope to pin 15 of the 7476. How many cycles are now displayed? By what factor did the first stage of counter divide the clock signal? On a graph like the one in Fig. 26-10(b), draw the observed waveform, showing its time relationship to the clock signal.

c. Move the vertical input of the oscilloscope to pin 11 of the 7476. How many cycles are displayed on the screen of the oscilloscope? Draw the displayed waveform in the appropriate place on the graph used in step 5b. Set the sweep controls of the oscilloscope back to their calibrated positions. Measure and record the period of this waveform. How does this compare to the period of the 555 timer signal?

d. A modulus-4 counter, when used for frequency division, can be called a _____ counter.

e. By what factor would a five-stage binary counter divide the clock signal?

ACTIVITY 26-3 IMPLEMENTING BOOLEAN EXPRESSIONS

Introduction

Completion of this activity will provide experience in implementing Boolean expressions using only NOR and NAND gates. Being limited to these gates forces one to use the universal property of NANDs and NORs. Use of this property can often reduce the number of chips, and the numbers of types of chips, needed in a logic system.

Supplies

(1) Power supply, 5-V dc
(1) Digital IC, 7400
(1) Digital IC, 7402
(4) Resistors, 220-Ω, $\frac{1}{4}$-W, 5%
(4) LED's, T-1 $\frac{3}{4}$, 40 mA maximum

Procedure

1. Using only one 7402 IC, design a logic circuit to directly implement the Boolean expression $X = \overline{(A + B) + C}$. *Hint:* Refer back to Fig. 26-4. Use a series-connected LED and resistor on the output and on each input to indicate the input and output logic levels. Just use a conductor to either V_{CC} or ground for switching the inputs to a logic 1 or a logic 0.

Construct the circuit you have designed and demonstrate its operation to your instructor. Turn in a neatly drawn logic diagram of your circuit. Include pin numbers of all gate leads. You need not show the LEDs and resistors on the inputs and output. Just mark the inputs A, B, and C; mark the output X.

2. Using one 7400 IC and one 7402 IC, design a logic circuit to directly implement $X = \overline{(AB + C)A}$. To aid you in designing your circuit, Fig. 26-11 shows the expression implemented with AND, OR, and NAND gates.

After constructing your circuit, demonstrate it to your instructor. Turn in the type of diagram described in step 1 above.

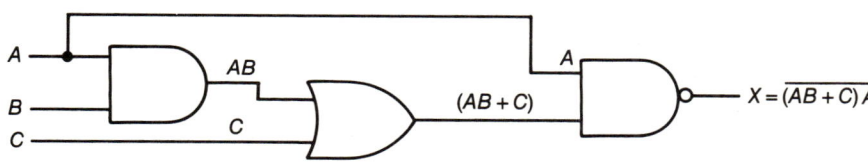

FIGURE 26-11 Circuit to implement $X = \overline{(AB + C) A}$.

Materials List

The following lists all equipment, components, supplies, and materials required to support the 26 chapters in this manual.

COMPONENT LIST

Transformers:

3 ea. dual, 12-V secondary and dual 115-V primary (Triad F-107Z or equivalent)

Resistors:

$\frac{1}{2}$-W, ±1%
1 ea., 6.49 Ω
1 ea., 71.5 Ω
1 ea., 100 kΩ

$\frac{1}{2}$-W, ±5%
1 ea., 1 Ω
1 ea., 4.7 Ω
2 ea., 10 Ω
1 ea., 22 Ω
1 ea., 27 Ω
1 ea., 33 Ω
1 ea., 43 Ω
1 ea., 100 Ω
4 ea., 220 Ω
1 ea., 300 Ω
1 ea., 330 Ω
1 ea., 560 Ω
1 ea., 820 Ω
3 ea., 1 kΩ
1 ea., 1.2 kΩ
1 ea., 1.5 kΩ
1 ea., 1.8 kΩ
2 ea., 2.2 kΩ
1 ea., 3 kΩ
1 ea., 3.3 kΩ
1 ea., 5.1 kΩ
2 ea., 10 kΩ
1 ea., 22 kΩ
2 ea., 68 kΩ
2 ea., 120 kΩ
2 ea., 150 kΩ
1 ea., 180 kΩ
2 ea., 910 kΩ

1-W, ±5%
1 ea., 100 Ω
1 ea., 220 Ω

2-W, ±5%
1 ea., 100 Ω

3-W, ±5%
1 ea., 125 Ω

5-W, ±5%
1 ea., 1000 Ω

20-W, ±5%
1 ea., 100 Ω

25-W, ±5%
1 ea., 10 Ω
1 ea., 50 Ω

50-W, ±5%
1 ea., 12 Ω

Potentiometers:

$\frac{1}{2}$-W, ±5%
1 ea., 1 kΩ
1 ea., 50 kΩ
1 ea., 1 MΩ

Capacitors:

10-V, ±10%, nonpolarized
1 ea., 100 μF

25-V, ±10%
2 ea., 0.01 μF
2 ea., 0.1 μF
2 ea., 10 μF

50-V, ±10%
1 ea., 0.05 μF
1 ea., 0.5 μF
1 ea., 1 μF
1 ea., 500 μF
1 ea., 1000 μF

Inductors:

1 ea., 1.0 mH, <10 Ω dc resistance, ±10%
1 ea., 2.5 mH, <10 Ω dc resistance, ±10%
1 ea., 0.4 H, 22 Ω, 275 mA

Switches:

3 ea., SPST (any current and voltage rating)
1 ea., SPST momentary contact

Diodes:

4 ea., 1N4002
1 ea., IN4733B, 5.6 V, 1 W, ±5%
4 ea., LED, T-1¾, 40 mA max.

Transistors:

1 ea., UJT 2N2646
1 ea., JFET (N channel) 2N5458
2 ea., NPN 2N4124
1 ea., PNP 2N4126
1 ea., MOSFET 2N3797

Integrated Circuits:

1 ea., 555
1 ea., 741
1 ea., 7400
1 ea., 7402
1 ea., 7476

Meters (panel)

1 ea., 100 µA, 650 Ω
1 ea., wattmeter, 75 W, 150 V, 1 A max.

Fuses:

3 ea., 1 A, with holders

Lamps:

3 ea., 14 V, 0.08 A, no. 756, with holders
1 ea., NE-2

Cells:

4 ea., alkaline D, with holders

Relay:

1 ea., 12-V dc

Magnetic Materials:

1 ea., magnetic compass
1 ea., assortment of steel rods
2 ea., bar magnets
2 ea., ferrite bars or rods

Miscellaneous Supplies:

Roll of tape
#28 magnet wire
#22 insulated wire (solid)
1 ea., index card, 3 × 5 in
1 ea., piece of sheet copper, 1 × 3 in (circuit board stock is acceptable)
1 ea., piece of sheet zinc, 1 × 3 in (galvanized steel is acceptable)
1 ea., steel wool pad
1 ea., dish of vinegar (standard 5% acetic acid content)

Equipment:

Variable ac supply, 0 to 120 V, 60 Hz
2 ea., Variable dc supply, 0 to 25 V
Oscilloscope with ×1 and ×10 probes
VOM
DMM
Soldering gun
Soldering pencil (23-W)
Ammeter, ac clamp-on

Appendix 1

Activity Report

Most of the activities in each chapter provide you with a detailed, step-by-step procedure that points out the major characteristics of the circuit being tested. The last activity for many chapters is quite different in that you must select most of the component values and determine all of the procedures to be used in proving experimentally that the circuit you designed performs as expected. Also, for these activities you will write a detailed report rather than answer specific questions as you collect the data. Your report must include at least the following:

An introduction explaining the purpose of the experiment or activity

A detailed list of the equipment you used in collecting data

An accurate schematic diagram of each circuit you constructed and tested

The calculations you used in arriving at component values

A detailed description of the test procedures you used in collecting data

A logical, organized presentation of the data collected

A list of the principles, laws, rules, etc., which are supported by the data presented

A discussion of the probable causes of disagreements between theoretical values and experimentally determined values

There is no one correct form for your report. Any format that is well-organized, neat, accurate, and complete is acceptable. The sample report that follows may give you some ideas to help you get started.

SAMPLE REPORT

Jane Doe   Oct. xx, 199x

ELECTRICITY 101

Report for Activity xx-x

Introduction
    The purpose of this activity was to design and test an unfiltered power supply which provides 4.3 V_{dc} with a 60-Hz ripple frequency. The supply was to be connected to a 20-Ω resistive load.

Design Procedure

Since a ripple frequency of 60 Hz was specified, half-wave rectification was used so that the supply could operate from a 60-Hz line. The complete schematic diagram of the tested circuit is shown in Fig. A-1. The component values were calculated as follows:

$$V_{dc} = \frac{0.637(V_P - V_D)}{2} \text{ rearranged, yields}$$

$$V_P = \frac{2V_{dc} + 0.637V_D}{0.637} = \frac{(2 \times 4.3\,V) + (0.637 \times 0.7\,V)}{0.637} = 14.2\,V$$

$$V_{rms} = 0.707 V_P = 0.707 \times 14.2\,V = 10\,V$$

$$I_{dc} = \frac{V_{dc}}{R_L} = \frac{4.3\,V}{20\,\Omega} = 0.215\,A$$

$$PIV \geq V_P \geq 14.2\,V$$

$$I_f \geq I_{dc} \geq 0.215\,A$$

$$P_{RL} = V_{RL} \times I_{RL} = V_{dc} \times I_{dc}$$

$$= 4.3\,V \times 0.215\,A = 0.925\,W$$

For safety reasons, the calculated value of P_{RL} was doubled and the nearest available power rating was selected.

Test Equipment Used
DMM, Fluke Model 8010 A
Oscilloscope/$\times 1$ probe, Tektronix Model xxx

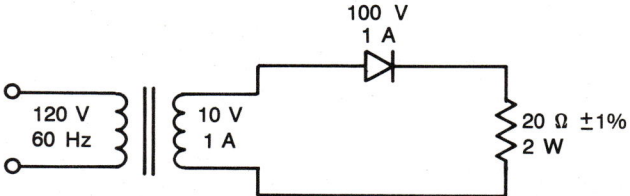

FIGURE A-1 Half-wave rectifier providing 4.3 V_{dc} across a 20-Ω load.

Test Procedure and Resulting Data

The circuit shown in Fig. A-1 was constructed and tested. The measured values of V_P and f in Table A-1 were obtained across the resistor using the oscilloscope with an $\times 1$ probe and calibrated horizontal and vertical amplifiers. The current entered in Table A-1 was obtained by inserting the DMM, on the dc current function, in series with the resistor in Fig. A-1. Finally, V_{dc} was measured with the DMM, on the dc voltage function. The experimentally determined frequency listed in Table A-1 was calculated using $f = 1/T$ and the value of T measured with the oscilloscope.

TABLE A-1 Expected and Measured Value for the Circuit in Fig. A-1

	Expected or Specified	Measured or Experimental
V_P	14.14 V	15.21 V
V_{rms}	10 V	10.76 V
f_{rip}	60 Hz	61.3 Hz
I_{dc}	0.215 A	0.236 A
V_{dc}	4.3 V	4.65 V
R_L	20 Ω	19.7 Ω

Formulas Supported by Data

The general agreement of the expected and experimental data in Table A-1 indicates that the standard formulas for half-wave rectifier circuits are appropriate for designing a simple, unfiltered power supply. These formulas are:

$$f_{rip} = f_{line}$$

$$V_{dc} = \frac{0.637(V_P - V_D)}{2}$$

$$I_{dc} = \frac{V_{dc}}{R_L}$$

$$PIV \simeq V_P$$

$$I_f \simeq I_{dc}$$

Discussion of Results

Inspection of Table A-1 shows that the measured values (except V_D and R_L) were larger than the expected values. This was probably the result of using a 1-A transformer rather than a 0.25-A transformer. At less than rated current, transformers usually produce a greater than rated secondary voltage. At light loads, the secondary voltage can be more than 5 percent above rated value.

Two factors worked together to make the measured values of I_{dc} larger than expected. One factor was the higher than expected value of secondary voltage, and the other was the lower than expected value of R_L.

Finally, it should be noted that test equipment tolerance could also account for some of the differences between expected and experimental values. However, the observed differences far exceed the errors that could be caused by the tolerances of the test equipment used in this activity. The greatest portion of the observed difference must, therefore, be attributed to the component tolerances and characteristics discussed above.

Appendix 2
BASIC Programs

The programs in this appendix will run under either BASICA or GWBASIC on an IBM or compatible personal computer. Only minor modifications would be required to operate these programs on other computers which have BASIC language capabilities.

Each program is followed by one or more sample runs to illustrate the required operator input and the resulting output. The sample runs should be especially useful for debugging programs which have been modified to operate on other than the IBM or compatible personal computer, or for tracking down errors made in entering the programs.

The programs are numbered to correspond to the appropriate chapters in the textbook. For example, Programs 19-1 and 19-2 are relevant to the materials and concepts presented in Chap. 19.

PROGRAM 2-1 OHM'S LAW

This program is based on Ohm's law and the power equation. The user may choose to enter any two variables and the program finds the remaining two.

```
10 CLS: SCREEN 0: WIDTH 80: CLEAR
20 PRINT TAB(37) "OHM'S LAW"
30 PRINT
40 PRINT "                    [CHOOSE ANY TWO VARIABLES]"
50 PRINT
60 PRINT"TYPE " CHR$(34) "V" CHR$(34) " FOR VOLTAGE"
70 PRINT"TYPE " CHR$(34) "I" CHR$(34) " FOR CURRENT"
80 PRINT"TYPE " CHR$(34) "R" CHR$(34) " FOR RESISTANCE"
90 PRINT"TYPE " CHR$(34) "P" CHR$(34) " FOR POWER"
100 PRINT:PRINT "               [USE ONLY NON-ZERO, POSITIVE UNIT VALUES]"
110 PRINT "Type"
120 Y$ = INKEY$: IF Y$="" THEN 120
130 IF Y$<>"V" AND Y$<>"I" AND Y$<>"R" AND Y$<>"P" AND Y$<>"v" AND Y$<>"i" AND Y$<>"r" AND Y$<>"p" THEN PRINT"WRONG KEY":GOTO 120
140 IF Y$="V" OR Y$="v" THEN INPUT "V = ";E
150 IF Y$="I" OR Y$="i" THEN INPUT "I = ";I
160 IF Y$="R" OR Y$="r" THEN INPUT "R = ";R
170 IF Y$="P" OR Y$="p" THEN INPUT "P = ";W
180 F=0 : G=G+1
190 IF E>0 THEN F=F+1
200 IF I>0 THEN F=F+1
210 IF R>0 THEN F=F+1
220 IF W>0 THEN F=F+1
230 IF F=0 AND G=1 THEN 270
240 IF F=2 GOTO 300
250 IF G=2 THEN 270
260 GOTO 110
270 PRINT "DO NOT ENTER ZERO OR A NEGATIVE NUMBER.":PRINT
280 PRINT "Press any key to continue."
290 X$=INKEY$: IF X$="" THEN 290 ELSE 10
300 IF E<>0 AND I<>0 THEN R=E/I:W=E*I
310 IF E<>0 AND R<>0 THEN I=E/R:W=E*I
320 IF E<>0 AND W<>0 THEN I=W/E:R=E/I
330 IF I<>0 AND R<>0 THEN E=I*R:W=E*I
340 IF I<>0 AND W<>0 THEN E=W/I:R=E/I
350 IF R<>0 AND W<>0 THEN E=SQR(R*W):I=E/R
360 PRINT
370 PRINT"VOLTAGE-------> ";E;" VOLTS"
380 PRINT"CURRENT-------> ";I;" AMPERES"
390 PRINT"RESISTANCE----> ";R;" OHMS"
400 PRINT"POWER--------> ";W;" WATTS"
410 CLEAR:PRINT:LOCATE 24,1
420 PRINT "DO ANOTHER? (Y/N)";
430 X$ = INKEY$:IF X$="" THEN 430
440 IF X$ <> "Y" AND X$ <> "y" AND X$ <> "N" AND X$ <>"n" THEN PRINT "WRONG KEY":GOTO 430
450 IF X$ = "Y" OR X$ ="y" THEN 10 ELSE END
```

SAMPLE RUN FOR OHM'S LAW

```
                              OHM'S LAW

                     [CHOOSE ANY TWO VARIABLES]

TYPE "V" FOR VOLTAGE
TYPE "I" FOR CURRENT
TYPE "R" FOR RESISTANCE
TYPE "P" FOR POWER

              [USE ONLY NON-ZERO, POSITIVE UNIT VALUES]
Type
V = ? 120
Type
R = ? 60

VOLTAGE------->    120  VOLTS
CURRENT------->    2    AMPERES
RESISTANCE---->    60   OHMS
POWER--------->    240  WATTS

DO ANOTHER? (Y/N)
```

PROGRAM 4-1 WIRE CALCULATOR

For many applications, this program eliminates the need for a wire table. It is menu-driven and allows the user a choice of three options.

```
10 CLS: SCREEN 0: WIDTH 80: CLEAR
20 PRINT TAB(35) "WIRE CALCULATOR": PRINT
30 PRINT TAB(15)"[1] FIND WIRE GAGE FROM RESISTANCE AND LENGTH
40 PRINT TAB(15)"[2] FIND WIRE GAGE FROM CURRENT (COPPER ONLY)
50 PRINT TAB(15)"[3] FIND WIRE RESISTANCE FROM GAGE AND LENGTH
60 PRINT: INPUT "ENTER YOUR CHOICE (1,2, OR 3)";C
70 IF C = 1 THEN 100
80 IF C = 2 THEN 210
90 IF C = 3 THEN 310 ELSE PRINT "WRONG KEY" : GOTO 60
100 INPUT "ENTER THE DESIRED RESISTANCE IN OHMS";R
110 INPUT "ENTER THE LENGTH OF THE WIRE IN FEET";L
120 PRINT "COPPER (C) OR ALUMINUM (A) WIRE?" :X$=""
130 M$ = INKEY$:IF M$ = "" THEN 130
140 IF M$ = "A" OR M$ = "a" THEN RO=17.34 :GOTO 160
150 IF M$ = "C" OR M$ = "c" THEN RO=10.369 ELSE PRINT "WRONG KEY":GOTO 130
160 A=RO*L/R
170 AWG=INT((LOG(A/105532!))/LOG(.79304)+.5)
180 GOSUB 470
190 PRINT:PRINT "THE AWG SIZE REQUIRED IS ";A$
200 GOTO 530
210 INPUT "ENTER THE CURRENT IN AMPERES";I
220 PRINT "IS THE WIRE GOING TO BE USED IN A COIL OR TRANSFORMER (Y/N)":X$=""
230 U$=INKEY$: IF U$="" THEN 230
240 IF U$ = "Y" OR U$ = "y" THEN C = 700 :GOTO 260
250 IF U$ = "N" OR U$ = "n" THEN C=274 ELSE PRINT "WRONG KEY":GOTO 220
260 A=C*I
270 AWG = INT(10*(LOG(325^2/A))/LOG(10)+.5)
280 GOSUB 470
290 PRINT:PRINT "THE AWG SIZE REQUIRED IS ";A$
300 GOTO 530
310 INPUT "ENTER THE AWG WIRE SIZE (0000 OR SMALLER)";AWG$
320 IF VAL(AWG$)>0 THEN AWG = VAL(AWG$):GOTO 380
330 IF AWG$ = "0" THEN AWG = 0:GOTO 380
340 IF AWG$ = "00" THEN AWG = -1:GOTO 380
350 IF AWG$ = "000" THEN AWG = -2:GOTO 380
360 IF AWG$ = "0000" THEN AWG = -3:GOTO 380
370 PRINT "INVALID GAGE":GOTO 310
380 INPUT "ENTER THE LENGTH OF THE WIRE IN FEET";L
390 PRINT "IS THE WIRE COPPER OR ALUMINUM (C/A)":X$=""
400 W$=INKEY$: IF W$="" THEN 400
410 IF W$ = "C" OR W$ = "c" THEN RH = 10.369 : GOTO 430
420 IF W$ = "A" OR W$ = "a" THEN RH = 17.34 ELSE PRINT "WRONG KEY":GOTO 390
430 A=105532!*.79304^AWG
440 R=RH*L/A
450 PRINT:PRINT "THE RESISTANCE OF THE WIRE IS";R;"OHMS"
460 GOTO 530
470 IF AWG => 0 THEN A$ = STR$(AWG)
480 IF AWG = -1 THEN A$ = "00"
490 IF AWG = -2 THEN A$ = "000"
500 IF AWG = -3 THEN A$ = "0000"
510 IF AWG < -3 THEN A$ = "LARGER THAN 0000"
520 RETURN
530 CLEAR:PRINT:LOCATE 24,1
540 PRINT "DO ANOTHER? (Y/N)";
550 X$ = INKEY$:IF X$="" THEN 550
560 IF X$ <> "Y" AND X$ <> "y" AND X$ <> "N" AND X$ <>"n" THEN PRINT "WRONG KEY"
:GOTO 550
570 IF X$ = "Y" OR X$ ="y" THEN 10 ELSE END
```

SAMPLE RUN FOR WIRE CALCULATOR

```
                        WIRE CALCULATOR

        [1] FIND WIRE GAGE FROM RESISTANCE AND LENGTH
        [2] FIND WIRE GAGE FROM CURRENT (COPPER ONLY)
        [3] FIND WIRE RESISTANCE FROM GAGE AND LENGTH
```

```
ENTER YOUR CHOICE (1,2, OR 3)? 3
ENTER THE AWG WIRE SIZE (0000 OR SMALLER)? 22
ENTER THE LENGTH OF THE WIRE IN FEET? 500
IS THE WIRE COPPER OR ALUMINUM (C/A)

THE RESISTANCE OF THE WIRE IS 8.069211 OHMS


DO ANOTHER? (Y/N)
```

PROGRAM 5-1 STANDARD VALUE FINDER

The values of many components, such as resistors, are spaced on a logarithmic scale. This program finds the nearest standard value and also states the error.

```
10 CLS: SCREEN 0: WIDTH 80: CLEAR
20 REM THIS PROGRAM HAS BEEN ADAPTED FROM ELECTRONIC DESIGN MAGAZINE, MAY 1974
30 REM THIS PROGRAM USES THE GEOMETRIC MEAN OF TWO STANDARD COMPONENT VALUES AS
THE DECISION POINT
40 PRINT "          SELECTING NEAREST STANDARD-VALUE RESISTOR": PRINT
50 INPUT"ENTER THE RESISTOR'S VALUE: ";X
60 IF X=0 OR X<0 THEN PRINT "Resistance must be >0.":GOTO 50
70 PRINT
80 INPUT"ENTER THE COMPONENT'S TOLERANCE (%): ";T
90 IF T>20 THEN PRINT "Tolerance can not be >20 %.":GOTO 80
100 A(4)=.0119927*INT(1+1.5*T+.004*T^2)
110 A(3)=INT(LOG(X)/LOG(10)-INT(2.2-3*A(4)))
120 XO=X
130 X=X/10^A(3)
140 FOR K = 1 TO 2
150 A(K)=INT(EXP(A(4)*(INT(LOG(X)/A(4))+K-1))+.5)
160 A(5)=.0000188*A(K)^3-.00335*A(K)^2+.164*A(K)-1.284
170 A(K)=A(K)+INT(A(5)*INT(3*A(4)+.8))
180 NEXT K
190 X=10^A(3)*A(INT(X/SQR(A(1)*A(2))+1))
200 X=INT(X*100+.5)/100
210 PRINT
220 PRINT"THE BEST STANDARD VALUE IS ";X
230 PRINT
240 E=INT((X-XO)/XO*1000+.5)/10
250 PRINT"     FOR A NOMINAL ERROR OF ";ABS(E);" %"
260 CLEAR:PRINT:LOCATE 24,1
270 PRINT "DO ANOTHER? (Y/N)";
280 X$ = INKEY$:IF X$="" THEN 280
290 IF X$ <> "Y" AND X$ <> "y" AND X$ <> "N" AND X$ <>"n" THEN PRINT "WRONG KEY"
:GOTO 280
300 IF X$ = "Y" OR X$ ="y" THEN 10 ELSE END
```

SAMPLE RUN FOR STANDARD VALUE FINDER

```
          SELECTING NEAREST STANDARD-VALUE RESISTOR

ENTER THE RESISTOR'S VALUE: ? 368

ENTER THE COMPONENT'S TOLERANCE (%): ? 1

THE BEST STANDARD VALUE IS  365

     FOR A NOMINAL ERROR OF  .8  %




DO ANOTHER? (Y/N)
```

PROGRAM 8-1 VOLTAGE DIVIDER DESIGNER

This program is useful for designing loaded voltage dividers that have a bleeder current.

```
10 CLS : SCREEN 0: WIDTH 80: CLEAR
20 PRINT "      VOLTAGE DIVIDER DESIGNER--ENTER LOAD VOLTAGES IN ASCENDING ORDER
": PRINT
30 INPUT "HOW MANY RESISTORS"; N
40 INPUT "WHAT IS THE SUPPLY VOLTAGE"; VT
50 INPUT "HOW MUCH BLEEDER CURRENT"; IB
60 IF IB = 0 THEN PRINT "YOU MUST HAVE A BLEEDER CURRENT": GOTO 50
70 FOR T = 1 TO (N - 1)
80 PRINT "ENTER VOLTAGE FOR LOAD #"; T;
90 INPUT V(T)
100 IF V(T) < V(T - 1) THEN PRINT : PRINT "LOAD VOLTAGES MUST BE ENTERED IN ASCE
NDING ORDER": GOTO 30
110 IF V(T) > VT THEN PRINT : PRINT "LOAD VOLTAGES MUST BE SMALLER THAN SUPPLY V
OLTAGE": GOTO 30
120 PRINT "ENTER CURRENT FOR LOAD #"; T;
130 INPUT I(T)
140 IT = IT + I(T)
150 NEXT T
160 V(T) = VT
170 FOR T = 1 TO N
180 IP = IP + I(T - 1)
190 VS = V(T) - V(T - 1)
200 R(T) = VS / (IB + IP)
210 P(T) = VS * (IB + IP)
220 NEXT T
230 PRINT : FOR T = 1 TO N
240 PRINT "RESISTOR "; T; " = "; R(T); " OHMS"
250 NEXT T
260 FOR T = 1 TO N
270 PRINT "RESISTOR "; T; " DISSIPATION = "; P(T); " WATTS"
280 NEXT T
290 IT = IT + IB
300 PRINT
310 PRINT "THE TOTAL DRAIN ON THE SUPPLY IS "; IT; " AMPS"
320 CLEAR : PRINT : LOCATE 24, 1
330 PRINT "DO ANOTHER? (Y/N)";
340 X$ = INKEY$: IF X$ = "" THEN 340
350 IF X$ <> "Y" AND X$ <> "y" AND X$ <> "N" AND X$ <> "n" THEN PRINT "WRONG KEY
": GOTO 340
360 IF X$ = "Y" OR X$ = "y" THEN 10 ELSE END
```

SAMPLE RUN FOR VOLTAGE DIVIDER DESIGNER

```
            VOLTAGE DIVIDER DESIGNER--ENTER LOAD VOLTAGES IN ASCENDING ORDER

HOW MANY RESISTORS? 4
WHAT IS THE SUPPLY VOLTAGE? 20
HOW MUCH BLEEDER CURRENT? .004
ENTER VOLTAGE FOR LOAD # 1 ? 5
ENTER CURRENT FOR LOAD # 1 ? .005
ENTER VOLTAGE FOR LOAD # 2 ? 12
ENTER CURRENT FOR LOAD # 2 ? .01
ENTER VOLTAGE FOR LOAD # 3 ? 15
ENTER CURRENT FOR LOAD # 3 ? .005

RESISTOR  1  =   1250      OHMS
RESISTOR  2  =   777.7778   OHMS
RESISTOR  3  =   157.8948   OHMS
RESISTOR  4  =   208.3333   OHMS
RESISTOR  1  DISSIPATION =   .02    WATTS
RESISTOR  2  DISSIPATION =   .063   WATTS
RESISTOR  3  DISSIPATION =   .057   WATTS
RESISTOR  4  DISSIPATION =   .12    WATTS

THE TOTAL DRAIN ON THE SUPPLY IS   .024   AMPS


DO ANOTHER? (Y/N)
```

PROGRAM 9-1 SIMULTANEOUS EQUATIONS

This program will solve up to 10 simultaneous equations. It is very helpful for solving dc networks using branch analysis, mesh analysis, etc.

```
10 CLS: SCREEN 0: WIDTH 80: CLEAR
20 PRINT TAB(30) "SIMULTANEOUS EQUATIONS": PRINT
30 INPUT "ENTER THE NUMBER OF EQUATIONS ";R:PRINT
40 FOR J = 1 TO R
50 PRINT "EQUATION ";J:
60 FOR I = 1 TO R+1
70 IF I = R+1 THEN 100
80 PRINT"   COEFFICIENT ";I;:PRINT" ";
90 GOTO 110
100 PRINT"   CONSTANT ";
110 INPUT A(J,I)
120 NEXT I:PRINT:NEXT J
130 FOR J = 1 TO R:FOR I = J TO R
140 IF A(I,J) <> 0 THEN 170
150 NEXT I
160 PRINT "NO UNIQUE SOLUTION": GOTO 370
170 FOR K = 1 TO R+1
180 X = A(J,K)
190 A(J,K) = A(I,K)
200 A(I,K) = X
210 NEXT K
220 Y = 1/A(J,J)
230 FOR K = 1 TO R+1
240 A(J,K) = Y*A(J,K)
250 NEXT K
260 FOR I = 1 TO R
270 IF I = J THEN 320
280 Y = -A(I,J)
290 FOR K = 1 TO R+1
300 A(I,K) = A(I,K)+Y*A(J,K)
310 NEXT K
320 NEXT I:NEXT J
330 PRINT
340 FOR I = 1 TO R
350 PRINT"X";I;" = ";:PRINT USING "##.##^^^^";A(I,R+1)
360 NEXT I
370 CLEAR:PRINT:LOCATE 24,1
380 PRINT "DO ANOTHER? (Y/N)";
390 X$ = INKEY$:IF X$="" THEN 390
400 IF X$ <> "Y" AND X$ <> "y" AND X$ <> "N" AND X$ <>"n" THEN PRINT "WRONG KEY":GOTO 390
410 IF X$ = "Y" OR X$ ="y" THEN 10 ELSE END
```

SAMPLE RUN FOR SIMULTANEOUS EQUATIONS

(the set of equations being solved is:)

$$3X + 2Y - Z = 0$$
$$X - Y + 2Z = -1$$
$$2X - 3Y + Z = -7$$

```
                    SIMULTANEOUS EQUATIONS

ENTER THE NUMBER OF EQUATIONS ? 3


EQUATION  1
   COEFFICIENT  1    ? 3
   COEFFICIENT  2    ? 2
   COEFFICIENT  3    ? -1
   CONSTANT ? 0

EQUATION  2
   COEFFICIENT  1    ? 1
   COEFFICIENT  2    ? -1
   COEFFICIENT  3    ? 2
   CONSTANT ? -1
```

```
EQUATION  3
  COEFFICIENT  1  ? 2
  COEFFICIENT  2  ? -3
  COEFFICIENT  3  ? 1
  CONSTANT ? -7

X 1  =  -1.00E+00
X 2  =   2.00E+00
X 3  =   1.00E+00

DO ANOTHER? (Y/N)
```

PROGRAM 9-2 WHEATSTONE BRIDGE

The Wheatstone bridge is a popular circuit. In many cases, a complete solution is not required. This program uses Thevenin's theorem to find the current and voltage drop for only the center resistor in the bridge. It is very easy to use because it "draws" the circuit on the screen.

```
10 CLS: SCREEN 0: WIDTH 80: CLEAR
20 PRINT"                       WHEATSTONE BRIDGE ANALYZER"
30 PRINT  TAB(5):FOR T=1 TO 17:PRINT".";:NEXT:PRINT
40 PRINT TAB(5):PRINT".";:PRINT TAB(20):PRINT". ."
50 PRINT TAB(5):PRINT".";:PRINT TAB(18):PRINT"R1    R2"
60 PRINT TAB(5):PRINT".";: PRINT TAB(17):PRINT".       ."
70 PRINT TAB(5):PRINT"V";:PRINT TAB(16):PRINT"....R5...."
80 PRINT  TAB(5):PRINT".";:PRINT TAB(17):PRINT".       ."
90 PRINT  TAB(5):PRINT".";:PRINT TAB(18):PRINT"R3    R4"
100 PRINT TAB(5):PRINT".";:PRINT TAB(20):PRINT". ."
110 PRINT TAB(5):FOR T=1 TO 17:PRINT".";:NEXT:PRINT
120 PRINT
130 FOR T=1 TO 5
140 PRINT"WHAT IS THE VALUE OF R";T;
150 INPUT R(T)
160 IF R(T)=<0 THEN PRINT "the resistance must be > 0.":GOTO 140
170 NEXT T
180 INPUT"WHAT IS THE VALUE OF V (THE SOURCE VOLTAGE)";V
190 X$=""
200 PRINT "IS THE SOURCE + OR - AT THE TOP OF THE BRIDGE?";
210 P$ = INKEY$:IF P$ = "" THEN 210
220 IF P$ = "+" OR P$ = "-" THEN 230 ELSE 200
230 PRINT
240 RTH=(R(1)*R(3)/(R(1)+R(3)))+(R(2)*R(4)/(R(2)+R(4)))
250 VTH=(R(1)*V/(R(1)+R(3)))-(R(2)*V/(R(2)+R(4)))
260 VL=VTH*R(5)/(RTH+R(5))
270 I=VL/R(5):PRINT
280 PRINT"THE CURRENT IN R5 IS ";ABS(I);" AMPS"
290 PRINT"THE DROP ACROSS R5 IS ";ABS(VL);" VOLTS"
300 IF I=0 THEN GOTO 320
310 IF I>0 AND P$ = "-" THEN PRINT "THE LEFT END OF R5 IS POSITIVE" ELSE PRINT " THE RIGHT END OF R5 IS POSITIVE"
320 CLEAR:PRINT:LOCATE 24,1
330 PRINT "DO ANOTHER? (Y/N)";
340 X$ = INKEY$:IF X$="" THEN 340
350 IF X$ <> "Y" AND X$ <> "y" AND X$ <> "N" AND X$ <>"n" THEN PRINT "WRONG KEY":GOTO 340
360 IF X$ = "Y" OR X$ ="y" THEN 10 ELSE END
```

SAMPLE RUN FOR WHEATSTONE BRIDGE

```
                    WHEATSTONE BRIDGE ANALYZER
     .................
     .             . .
     .            R1  R2
     .           .       .
     V           ....R5....
     .           .       .
     .            R3  R4
     .             . .
     .................
```

```
WHAT IS THE VALUE OF R 1 ? 10
WHAT IS THE VALUE OF R 2 ? 30
WHAT IS THE VALUE OF R 3 ? 20
WHAT IS THE VALUE OF R 4 ? 40
WHAT IS THE VALUE OF R 5 ? 50
WHAT IS THE VALUE OF V (THE SOURCE VOLTAGE)? 10
IS THE SOURCE + OR - AT THE TOP OF THE BRIDGE?

THE CURRENT IN R5 IS  1.290323E-02  AMPS
THE DROP ACROSS R5 IS  .6451613  VOLTS
THE RIGHT END OF R5 IS POSITIVE

DO ANOTHER? (Y/N)
```

PROGRAM 9-3   DC NETWORK ANALYSIS

The well-known circuit analysis packages such as SPICE and ECAP use nodal analysis. This program is no exception and can provide a detailed analysis for complex dc networks such as the one shown below, which is the same circuit analyzed in the sample run.

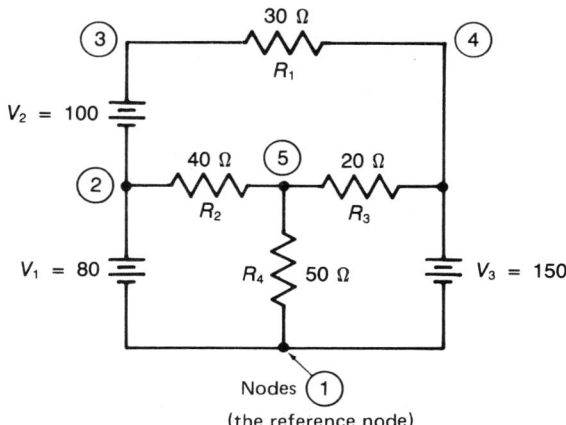

```
10 CLS: SCREEN 0: WIDTH 80: CLEAR
20 REM WRITTEN BY PROFESSOR J. LONEY, CALIFORNIA UNIVERSITY OF PA
30 PRINT TAB(30) "DC NETWORK ANALYSIS"
40 PRINT TAB(22) "(Don't use zero for a node number)"
50 PRINT TAB(23) "(The reference node is number 1)": PRINT
60 OPTION BASE 1: DIM GA(10,20),R(20,10,10)
70 INPUT "ENTER THE NUMBER OF NODES";N
80 INPUT "ENTER THE NUMBER OF RESISTORS";R
90 FOR K = 1 TO R
100 PRINT "ENTER THE VALUE FOR R";K;"IN OHMS";:INPUT X
110 PRINT"ENTER THE TWO NODES CONNECTED TO R";K;:INPUT L,M
120 R(K,L,M) = X
130 G(L,M) = (-1/X) + G(L,M)+G(M,L):G(M,L) = 0
140 NEXT K
150 INPUT "ENTER THE NUMBER OF SOURCES";S
160 FOR K = 1 TO S
170 PRINT"ENTER THE VALUE OF SOURCE";K;"IN VOLTS";:INPUT X
180 INPUT "ENTER THE POSITIVE NODE";M
190 INPUT "ENTER THE NEGATIVE NODE";L
200 F(L,M) = X*1000
210 G(L,M) = G(L,M) - 1000+G(M,L)
220 NEXT K
230 FOR I = 2 TO N:FOR J = 1 TO N
240 IF I >< J THEN G(I,I) = G(I,I) - G(I,J) - G(J,I)
250 NEXT J:NEXT I
260 FOR I = 1 TO N:FOR J = 1 TO N
270 IF I = J THEN 290
280 IF G(I,J)-G(J,I)>0 THEN G(I,J) = G(J,I) ELSE G(J,I) = G(I,J)
290 NEXT J:NEXT I
300 FOR I = 2 TO N:FOR J = 2 TO N
310 GA(I,J) = G(I,J)
320 GA(I,N+J) = 0:NEXT J
330 GA(I,N+I) = 1:NEXT I
340 FOR I = 2 TO N
```

```
350 A = GA(I,I)
360 FOR J = I TO 2*N
370 GA(I,J) = GA(I,J)/A:NEXT J
380 FOR K = 2 TO N
390 IF (K-I) = 0 THEN 440
400 B = GA(K,I)
410 FOR J = I TO 2*N
420 GA(K,J) = GA(K,J)-B*GA(I,J)
430 NEXT J
440 NEXT K:NEXT I
450 FOR J = 2 TO N
460 JN = J+N
470 FOR I = 2 TO N
480 GI(I,J) = GA(I,JN)
490 NEXT I:NEXT J
500 FOR J = 1 TO N:FOR K = 1 TO N
510 IF I >< J THEN C(J) = C(J) + F(K,J) - F(J,K)
520 NEXT K:NEXT J:PRINT
530 FOR J = 2 TO N:FOR K = 2 TO N
540 V(J)= V(J) + GI(J,K)*C(K):NEXT K
550 PRINT"V(";J;") = ";
560 PRINT USING "##.##^^^^";V(J);:PRINT " VOLTS"
570 NEXT J:PRINT
580 PRINT "PRESS ANY KEY TO CONTINUE": PRINT
590 X$=INKEY$:IF X$="" THEN 590
600 FOR I = 1 TO R:FOR J = 1 TO N:FOR K = 1 TO N
610 IF R(I,J,K) = 0 THEN 650
620 E1 = (V(J)-V(K))/R(I,J,K)
630 PRINT"THE CURRENT THROUGH R";I;"IS";
640 PRINT USING "##.##^^^^";ABS(E1);:PRINT " AMPERES"
650 NEXT K:NEXT J:NEXT I
660 CLEAR:PRINT:LOCATE 24,1
670 PRINT "DO ANOTHER? (Y/N)"
680 X$ = INKEY$:IF X$="" THEN 680
690 IF X$ <> "Y" AND X$ <> "y" AND X$ <> "N" AND X$ <>"n" THEN PRINT "WRONG KEY"
:GOTO 680
700 IF X$ = "Y" OR X$ ="y" THEN 10 ELSE END
```

SAMPLE RUN FOR DC NETWORK ANALYSIS

```
                    DC NETWORK ANALYSIS
            (Don't use zero for a node number)
              (The reference node is number 1)

ENTER THE NUMBER OF NODES? 5
ENTER THE NUMBER OF RESISTORS? 4
ENTER THE VALUE FOR R 1 IN OHMS? 30
ENTER THE TWO NODES CONNECTED TO R 1 ? 3,4
ENTER THE VALUE FOR R 2 IN OHMS? 40
ENTER THE TWO NODES CONNECTED TO R 2 ? 2,5
ENTER THE VALUE FOR R 3 IN OHMS? 20
ENTER THE TWO NODES CONNECTED TO R 3 ? 5,4
ENTER THE VALUE FOR R 4 IN OHMS? 50
ENTER THE TWO NODES CONNECTED TO R 4 ? 1,5
ENTER THE NUMBER OF SOURCES? 3
ENTER THE VALUE OF SOURCE 1 IN VOLTS? 80
ENTER THE POSITIVE NODE? 2
ENTER THE NEGATIVE NODE? 1
ENTER THE VALUE OF SOURCE 2 IN VOLTS? 100
ENTER THE POSITIVE NODE? 3
ENTER THE NEGATIVE NODE? 2
ENTER THE VALUE OF SOURCE 3 IN VOLTS? 150
ENTER THE POSITIVE NODE? 4
ENTER THE NEGATIVE NODE? 1


V( 2 )  =   8.00E+01 VOLTS
V( 3 )  =   1.80E+02 VOLTS
V( 4 )  =   1.50E+02 VOLTS
V( 5 )  =   1.00E+02 VOLTS

PRESS ANY KEY TO CONTINUE

THE CURRENT THROUGH R 1 IS 1.00E+00 AMPERES
THE CURRENT THROUGH R 2 IS 5.00E-01 AMPERES
```

THE CURRENT THROUGH R 3 IS 2.50E+00 AMPERES
THE CURRENT THROUGH R 4 IS 2.00E+00 AMPERES

DO ANOTHER? (Y/N)

PROGRAM 14-1 INDUCTIVE REACTANCE

Inductance, frequency, or inductive reactance can be determined by this program when any two of the three variables are specified. All data are entered in base units, and all output is printed in base units.

```
10 CLS: SCREEN 0: WIDTH 80: CLEAR
20 PRINT TAB(35) "INDUCTIVE REACTANCE":PRINT
30 PRINT
40 INPUT "Do you want to determine inductance (1), frequency (2) or inductive
         reactance (3)"; Q
50 IF Q <1 OR Q>3 THEN 40
60 IF Q <> 1 THEN INPUT "The inductance, in henrys, is";L
70 IF Q <> 2 THEN INPUT "The frequency, in hertz, is"; F
80 IF Q <> 3 THEN INPUT "The inductive reactance, in ohms, is";XL
90 PRINT:PRINT
100 IF Q = 1 THEN L=XL/(6.2832*F): PRINT "The inductance is"; L; "henrys."
110 IF Q = 2 THEN F=XL/(6.2832*L): PRINT "The frequency is"; F; "hertz."
120 IF Q = 3 THEN XL=6.2832*F*L: PRINT "The inductive reactance is"; XL; "ohms."
130 CLEAR:PRINT:LOCATE 24,1
140 PRINT "DO ANOTHER? (Y/N)";
150 X$ = INKEY$:IF X$="" THEN 150
160 IF X$ <> "Y" AND X$ <> "y" AND X$ <> "N" AND X$ <>"n" THEN PRINT "WRONG KEY"
    :GOTO 150
170 IF X$ = "Y" OR X$ ="y" THEN 10 ELSE END
```

SAMPLE RUNS FOR INDUCTIVE REACTANCE

```
                              INDUCTIVE REACTANCE

Do you want to determine inductance (1), frequency (2) or inductive
reactance (3)? 3
The inductance, in henrys, is? .002
The frequency, in hertz, is? 23e3


The inductive reactance is 289.0272 ohms.
```

DO ANOTHER? (Y/N)

```
                              INDUCTIVE REACTANCE

Do you want to determine inductance (1), frequency (2) or inductive
reactance (3)? 2
The inductance, in henrys, is? .005
The inductive reactance, in ohms, is? 820


The frequency is 26101.35 hertz.
```

DO ANOTHER? (Y/N)

PROGRAM 14-2 AC INDUCTOR CIRCUITS

This program calculates all currents, voltages, and inductive reactances for any inductor circuit except a series-parallel circuit. The operator supplies the values for the frequency, the source voltage, and the individual inductors.

So that all the inputs and the outputs will appear on the screen, the number of inductors has been limited to 4. The limit can be changed by changing both 4's in line 80 to any desired number.

```
10 CLS: SCREEN 0: WIDTH 80: CLEAR
20 PRINT TAB(35) "AC INDUCTOR CIRCUITS":PRINT
30  PRINT "This program calculates all currents, voltages and reactances for
            single, series or parallel inductor circuits.": PRINT
40  INPUT "Series (1), parallel (2), or single (3)"; Q
50   IF Q<1 OR Q>3 OR Q<> INT(Q) THEN 40
60  IF Q=3 THEN N=1: J=1: GOTO 90
70  INPUT "How many inductors"; N
80  IF N<1 OR N>4 OR N<> INT (N) THEN PRINT "Enter an integer between 1 and 4":
     GOTO 70
90  INPUT "The source voltage, in volts, is";V
100 INPUT "The frequency, in hertz, is";F
110 IF Q=3 THEN INPUT "The inductance, in henrys, is";L(J)
120 FOR J = 1 TO N
130    IF Q<3 THEN PRINT "Inductance number ";J;", in henrys, is";:
       INPUT L(J)
140    X(J) = 6.2832*F*L(J)
150    IF Q=1 THEN XT=XT+X(J): LT=LT+L(J) ELSE LP=LP+1/L(J): XP=XP+1/X(J)
160    NEXT J
170 IF Q>1 THEN LT=1/LP: XT=1/XP
180 IT=V/XT: PRINT
190 FOR J=1 TO N
200    IF Q<3 THEN PRINT "XL";J;"=";X(J);"ohms."
210    NEXT J
220 IF Q<3 THEN PRINT "XLt = ";XT;"ohms." ELSE PRINT "XL = ";XT;"ohms."
230 IF Q<3 THEN PRINT "Lt = ";LT;"henrys."
240 IF Q<3 THEN PRINT "It = ";IT*1000; "milliamperes." ELSE
     PRINT "I = ";IT*1000;"milliamperes."
250 IF Q=3 GOTO 290
260 FOR J=1 TO N
270    IF Q=1 THEN PRINT "VL";J;"= "; IT*X(J);"volts." ELSE
       PRINT "IL";J;"= ";1000*V/X(J); "milliamperes."
280    NEXT J
290 CLEAR:PRINT:LOCATE 24,1
300 PRINT "DO ANOTHER? (Y/N)";
310 X$ = INKEY$:IF X$="" THEN 310
320 IF X$ <> "Y" AND X$ <> "y" AND X$ <> "N" AND X$ <>"n" THEN PRINT "WRONG KEY"
    :GOTO 310
330 IF X$ = "Y" OR X$ ="y" THEN 10 ELSE END
```

SAMPLE RUNS FOR AC INDUCTOR CIRCUITS

```
                      AC INDUCTOR CIRCUITS

This program calculates all currents, voltages and reactances for
single, series or parallel inductor circuits.

Series (1), parallel (2), or single (3)? 2
How many inductors? 3
The source voltage, in volts, is? 42
The frequency, in hertz, is? 1000
Inductance number  1 , in henrys, is? .03
Inductance number  2 , in henrys, is? .027
Inductance number  3 , in henrys, is? .05

XL 1 = 188.496 ohms.
XL 2 = 169.6464 ohms.
XL 3 = 314.16 ohms.
XLt =  69.52721 ohms.
Lt =   1.106557E-02 henrys.
It =   604.0801 milliamperes.
IL 1 =  222.8164 milliamperes.
IL 2 =  247.5738 milliamperes.
```

```
IL 3 =   133.6899 milliamperes.

DO ANOTHER? (Y/N)
```

```
                          AC INDUCTOR CIRCUITS

This program calculates all currents, voltages and reactances for
single, series or parallel inductor circuits.

Series (1), parallel (2), or single (3)? 3
The source voltage, in volts, is? 15
The frequency, in hertz, is? 1200
The inductance, in henrys, is? 50e-3

XL =  376.992 ohms.
I  =  39.78864 milliamperes.


DO ANOTHER? (Y/N)
```

PROGRAM 15-1   SERIES CAPACITORS

This program will either determine the equivalent capacitance of up to five series capacitors or determine the value of capacitor needed in series with a specified capacitor to provide a specified equivalent capacitance. With either of these options, the program also provides the voltage distribution for the series capacitors. The voltage distribution is given as a portion of the source voltage.

```
10 CLS: SCREEN 0: WIDTH 80: CLEAR
20 PRINT TAB(35) "SERIES CAPACITORS": PRINT
30 PRINT "Enter all capacitances in microfarads.": PRINT
40 PRINT "Two options are provided:
            1. determine the equivalent capacitance of up to five
               series capacitances."
50 PRINT "    2. determine the capacitance required in series with a
               specified capacitance to provide a specified
               equivalent capacitance.": PRINT
60 INPUT "Enter the number of the option you desire. ", O
70 IF O = 1 OR O = 2 THEN ON O GOTO 100, 280
80 PRINT "Please select one of the two options."
90 GOTO 60
100 PRINT
110 INPUT "How many capacitors are in series ";N
120 IF N>=1 AND N<=5 AND N=INT(N) THEN 150
130   PRINT "Must be an integer from one to five"
140 GOTO 100
150 FOR I=1 TO N
160 PRINT "     Capacitance number"; I; "is";
170 INPUT X
180 C = C + 1/X
190 C(I) = X
200 NEXT I
210 C = 1/C : PRINT
220 PRINT "Total (equivalent) capacitance is " C "microfarads"
230 FOR I=1 TO N
240 VC = C/C(I)
250 PRINT "Voltage across C"I" equals"VC" times the source voltage."
260 NEXT I
270 GOTO 410
280 PRINT
290 INPUT "Desired equivalent capacitance (CT)"; CE
300 INPUT "Capacitance to be in series (C1)"; C1
310 IF C1>CE THEN 340
320 PRINT "C1 must be larger than CT" : PRINT
330 GOTO 300
340 C2=CE*C1/(C1-CE)
350 PRINT
360 PRINT "The capacitance needed in series is" C2 "microfarads"
370 VC1 = CE/C1
380 VC2 = CE/C2
390 PRINT "Voltage across C1 equals "VC1"times the source voltage."
400 PRINT "Voltage across C2 equals "VC2"times the source voltage."
410 CLEAR:PRINT:LOCATE 24,1
```

```
420 PRINT "DO ANOTHER? (Y/N)";
430 X$ = INKEY$:IF X$="" THEN 430
440 IF X$ <> "Y" AND X$ <> "y" AND X$ <> "N" AND X$ <>"n" THEN PRINT "WRONG KEY"
:GOTO 430
450 IF X$ = "Y" OR X$ ="y" THEN 10 ELSE END
```

SAMPLE RUNS FOR SERIES CAPACITORS

```
                         SERIES CAPACITORS

Enter all capacitances in microfarads.

Two options are provided:
     1. determine the equivalent capacitance of up to five
        series capacitances.
     2. determine the capacitance required in series with a
        specified capacitance to provide a specified
        equivalent capacitance.

Enter the number of the option you desire. 1

How many capacitors are in series ? 3
     Capacitance number 1 is? .01
     Capacitance number 2 is? .033
     Capacitance number 3 is? .015

Total (equivalent) capacitance is  5.076923E-03 microfarads
Voltage across C 1    equals .5076923   times the source voltage.
Voltage across C 2    equals .1538462   times the source voltage.
Voltage across C 3    equals .3384616   times the source voltage.

DO ANOTHER? (Y/N)


                         SERIES CAPACITORS

Enter all capacitances in microfarads.

Two options are provided:
     1. determine the equivalent capacitance of up to five
        series capacitances.
     2. determine the capacitance required in series with a
        specified capacitance to provide a specified
        equivalent capacitance.

Enter the number of the option you desire. 2

Desired equivalent capacitance (CT)? .156
Capacitance to be in series (C1)? .39

The capacitance needed in series is .26 microfarads
Voltage across C1 equals   .4 times the source voltage.
Voltage across C2 equals   .6 times the source voltage.

DO ANOTHER? (Y/N)
```

PROGRAM 16-1 *RC* TIME CONSTANTS

This program will determine any one of the variables involved in an *RC* time constant circuit. The variables can be determined for either a charging or a discharging capacitor.

The program always asks for the value of the source voltage. If no source is involved, like when a charged capacitor is discharging through a resistor, just enter 0 for the source voltage. Likewise enter 0 for either the starting or the final voltage on the capacitor when the capacitor is charging from 0 V or discharging to 0 V, respectively.

This program will also work when the starting voltage on the capacitor is series-aiding the source voltage. The output will indicate whether the capacitor is still discharging at the end of the allotted time or whether it has finished discharging and is recharging back toward the source voltage. Of course, when the capacitor voltage is series-aiding the source voltage, the capacitor's voltage must be entered as a negative value.

For this program, a capacitor is assumed to be completely charged or discharged after 7 time constants. To change the assumption from 7τ to 5τ, change the (-7) and the LN = 7 in line 300 to (-5) and LN = 5, respectively.

```
10  CLS: SCREEN 0: WIDTH 80: CLEAR
20  DIM A$(2): A$(1)="dis": A$(2)="discharge and re"
30  PRINT TAB(30) "RC TIME CONSTANTS": PRINT
40  PRINT "This program determines any of the variables in a simple RC"
50  PRINT "time constant circuit.": PRINT
60  PRINT "The options are to determine:"
70  PRINT "    1. t (time to reach a specified capacitor voltage.)"
80  PRINT "    2. R (resistance required for a specified time.)"
90  PRINT "    3. C (capacitance required for a specified time.)"
100 PRINT "    4. Vct (capacitor voltage after a specified time.)"
110 PRINT: INPUT "Which option ";N
120    IF N<1 OR N>4 OR N<>INT(N) THEN 110
130 PRINT: INPUT "The voltage (in volts) already on the capacitor is";VCE
140 INPUT "The source voltage (in volts) is";E
150    IF E=VCE THEN PRINT: PRINT "Equilibrium exists!": GOTO 340
160 IF N=1 THEN 190
170    INPUT "The time (in seconds) will be";T
180    IF T<0 THEN PRINT: PRINT "Time must be non-negative!": GOTO 170
190 IF N=2 THEN 220
200    INPUT "The resistance (in megohms) is";R
210    IF R<=0 THEN PRINT: PRINT "Resistance must be positive!": GOTO 200
220 IF N=3 THEN 250
230    INPUT "The capacitance (in microfarads) is";C
240    IF C<=0 THEN PRINT: PRINT "Capacitance must be positive!": GOTO 230
250 IF N=4 THEN VCT=E-(E-VCE)*EXP(-T/(R*C)): GOTO 310
260    INPUT "The final voltage (in volts) on the capacitor will be";VCT
270    IF (VCE-VCT)*(E-VCT)<=0 THEN 300
280       PRINT: PRINT "The final voltage must be between the initial"
290       PRINT "voltage and the source voltage!": GOTO 340
300    IF (E-VCT)/(E-VCE)<EXP(-7) THEN LN=7 ELSE LN=LOG((E-VCE)/(E-VCT))
310 IF VCE*VCT<0 THEN Q=2 ELSE Q=-(ABS(VCE)>ABS(VCT))
320 PRINT: PRINT
330 ON N GOSUB 350,380,410,440
340 GOTO 470
350 PRINT "The time required to ";A$(Q);"charge the";C;"microfarad capacitor ";
360    PRINT "to";VCT;"volts through a";R;"megohm resistor is";R*C*LN;"seconds."
370    RETURN
380 PRINT "The required resistance to ";A$(Q);"charge the";C;
390    PRINT "microfarad capacitor to";VCT;"volts in";T;"seconds is";T/(C*LN);
400    PRINT "megohms.": RETURN
410 PRINT "The required capacitance to ";A$(Q);"charge through the";R;
420    PRINT "megohm resistor to";VCT;"volts in";T;"seconds is";T/(R*LN);
430    PRINT "microfarads.": RETURN
440 PRINT "The";C;"microfarad capacitor's voltage after a ";A$(Q);
450    PRINT "charge time of";T;"seconds through a";R;"megohm resistor is";
460    PRINT VCT;"volts.": RETURN
470 CLEAR:PRINT:LOCATE 24,1
480 PRINT "DO ANOTHER? (Y/N)";
490 X$ = INKEY$:IF X$="" THEN 490
500 IF X$ <> "Y" AND X$ <> "y" AND X$ <> "N" AND X$ <>"n" THEN PRINT "WRONG KEY"
    :GOTO 490
510 IF X$ = "Y" OR X$ ="y" THEN 10 ELSE END
```

SAMPLE RUN FOR *RC* TIME CONSTANTS

```
                         RC TIME CONSTANTS

This program determines any of the variables in a simple RC
time constant circuit.

The options are to determine:
    1. t (time to reach a specified capacitor voltage.)
    2. R (resistance required for a specified time.)
    3. C (capacitance required for a specified time.)
    4. Vct (capacitor voltage after a specified time.)
```

```
Which option ? 1
   The voltage (in volts) already on the capacitor is? 10
   The source voltage (in volts) is? 40
   The resistance (in megohms) is? 2
   The capacitance (in microfarads) is? .1
   The final voltage (in volts) on the capacitor will be? 25

The time required to charge the .1 microfarad capacitor to 25 volts through a 2
megohm resistor is .1386294 seconds.

DO ANOTHER? (Y/N)
```

PROGRAM 16-2 CAPACITIVE REACTANCE

This program determines capacitive reactance, frequency, or capacitance where two of the three variables are known. Capacitance is both entered and printed out in microfarads. Other variables are handled in base units.

```
10 CLS: SCREEN 0: WIDTH 80: CLEAR
20 PRINT TAB(35) "CAPACITIVE REACTANCE": PRINT
30 PI = 3.1416
40 INPUT "Do you want to determine capacitance (1), frequency (2) or capacitive
        reactance (3)"; OP
50 IF OP <1 OR OP >3 THEN 40 ELSE PRINT
60 IF OP <> 1 THEN INPUT "The capacitance, in microfarads, is"; C
70 IF OP <> 2 THEN INPUT "The frequency, in hertz, is"; F
80 IF OP <> 3 THEN INPUT "The capacitive reactance, in ohms, is";XC
90 PRINT:PRINT
100 IF OP = 1 THEN C=1000000!/(2*PI*F*XC): PRINT "The capacitance is"; C; "micro
farads."
110 IF OP = 2 THEN F=1000000!/(2*PI*XC*C): PRINT "The frequency is"; F; "hertz."
120 IF OP = 3 THEN XC=1000000!/(2*PI*F*C): PRINT "The capacitive reactance is";
XC; "ohms."
130 CLEAR:PRINT:LOCATE 24,1
140 PRINT "DO ANOTHER? (Y/N)";
150 X$ = INKEY$:IF X$="" THEN 150
160 IF X$ <> "Y" AND X$ <> "y" AND X$ <> "N" AND X$ <>"n" THEN PRINT "WRONG KEY"
:GOTO 150
170 IF X$ = "Y" OR X$ ="y" THEN 10 ELSE END
```

SAMPLE RUNS FOR CAPACITIVE REACTANCE

```
                            CAPACITIVE REACTANCE

Do you want to determine capacitance (1), frequency (2) or capacitive
reactance (3)? 1

The frequency, in hertz, is? 2.5e3
The capacitive reactance, in ohms, is? 1250


The capacitance is 5.092947E-02 microfarads.

DO ANOTHER? (Y/N)


                            CAPACITIVE REACTANCE

Do you want to determine capacitance (1), frequency (2) or capacitive
reactance (3)? 3

The capacitance, in microfarads, is? 220e-6
The frequency, in hertz, is? 456e3


The capacitive reactance is 1586.469 ohms.

DO ANOTHER? (Y/N)
```

PROGRAM 16-3 AC CAPACITOR CIRCUITS

This program solves for all currents, voltages, and capacitive reactances for single, series, or parallel capacitor circuits. It is designed to accommodate a maximum of four capacitors; however, this maximum can be changed to any number desired by changing the 4 in line 80 to the number desired.

```
10 CLS: SCREEN 0: WIDTH 80: CLEAR
20 PRINT TAB(30) "AC CAPACITOR CIRCUITS":PRINT
30  PRINT "This program calculates all currents, voltages and reactances for
          single, series or parallel capacitor circuits.": PRINT
40  INPUT "Series (1), parallel (2), or single (3)"; Q
50  IF Q<1 OR Q>3 OR Q<> INT(Q) THEN 40 ELSE PRINT
60 IF Q>2 THEN N=1: J=1: GOTO 90
70 INPUT "How many capacitors"; N
80 IF N<1 OR N>4 OR N<> INT (N) THEN PRINT "Enter an integer between 1 and 4":
   GOTO 70
90 INPUT "The source voltage, in volts, is";V
100 INPUT "The frequency, in hertz, is";F
110 IF Q>2 THEN INPUT "The capacitance, in microfarads, is";C(J)
120 FOR J = 1 TO N
130    IF Q<3 THEN PRINT "Capacitance number ";J;", in microfarads, is";:
       INPUT C(J)
140    X(J) = 159155!/(F*C(J))
150    IF Q=1 THEN XT=XT+X(J): CP=CP+1/C(J) ELSE CT=CT+C(J): XP=XP+1/X(J)
160    NEXT J
170 IF Q=1 THEN CT=1/CP: ELSE XT=1/XP
180 IT=V/XT: PRINT
190 FOR J=1 TO N
200    IF Q<3 THEN PRINT "Xc";J;"=";X(J);"ohms."
210    NEXT J
220 IF Q<3 THEN PRINT "Xct = ";XT;"ohms." ELSE PRINT "Xc = ";XT;"ohms."
230 IF Q<3 THEN PRINT "Ct = ";CT;"microfarads."
240 IF Q<3 THEN PRINT "It = ";IT*1000; "milliamperes." ELSE
    PRINT "I = ";IT*1000;"milliamperes."
250 IF Q>2 GOTO 290
260 FOR J=1 TO N
270    IF Q=1 THEN PRINT "VC";J;"= "; IT*X(J);"volts." ELSE
       PRINT "IC";J;"= ";1000*V/X(J); "milliamperes."
280    NEXT J
290 CLEAR:PRINT:LOCATE 24,1
300 PRINT "DO ANOTHER? (Y/N)";
310 X$ = INKEY$:IF X$="" THEN 310
320 IF X$ <> "Y" AND X$ <> "y" AND X$ <> "N" AND X$ <>"n" THEN PRINT "WRONG KEY"
    :GOTO 310
330 IF X$ = "Y" OR X$ ="y" THEN 10 ELSE END
```

SAMPLE RUNS FOR AC CAPACITOR CIRCUITS

```
                         AC CAPACITOR CIRCUITS

This program calculates all currents, voltages and reactances for
single, series or parallel capacitor circuits.

Series (1), parallel (2), or single (3)? 1

How many capacitors? 2
The source voltage, in volts, is? 36
The frequency, in hertz, is? 1800
Capacitance number  1 , in microfarads, is? .01
Capacitance number  2 , in microfarads, is? .033

Xc 1 = 8841.944 ohms.
Xc 2 = 2679.377 ohms.
Xct =  11521.32 ohms.
Ct =   7.674419E-03 microfarads.
It =   3.124642 milliamperes.
VC 1 =  27.62791 volts.
VC 2 =   8.372093 volts.



DO ANOTHER? (Y/N)
```

AC CAPACITOR CIRCUITS

This program calculates all currents, voltages and reactances for single, series or parallel capacitor circuits.

Series (1), parallel (2), or single (3)? 2

How many capacitors? 2
The source voltage, in volts, is? 45
The frequency, in hertz, is? 4500
Capacitance number 1 , in microfarads, is? .002
Capacitance number 2 , in microfarads, is? .005

Xc 1 = 17683.89 ohms.
Xc 2 = 7073.556 ohms.
Xct = 5052.54 ohms.
Ct =   .007 microfarads.
It =   8.906412 milliamperes.
IC 1 = 2.544689 milliamperes.
IC 2 = 6.361723 milliamperes.

DO ANOTHER? (Y/N)

PROGRAM 17-1 *RCL* CIRCUITS

This program solves any series or parallel *RC*, *RL*, or *RCL* circuit. It will calculate as many unknown values as possible from the information provided about the circuit. For example, if the operator provides only the resistance and the source voltage for a parallel *RCL* circuit, the program will print out the values of I_R and P. Thus, the program can be used for simple Ohm's law problems involving only one resistance, capacitance, inductance, or reactance.

The program does not check to determine if conflicting data have been entered by the operator. For example, if the operator enters 10 V, 1000 Ω, 100 Hz, and 3 H for V_T, X_L, f, and L, respectively, the program will use these values, as needed, to calculate other values even though 3 H of inductance does not have 1000 Ω of reactance at 100 Hz. In other words, "garbage in . . . garbage out."

```
10  CLS: SCREEN 0: WIDTH 80: CLEAR
20  PRINT TAB(35) "RCL CIRCUITS":PRINT
30   DIM S$(17), U$(17), C(17), I(17), E(17) :A=1
40  FOR J=1 TO 17: READ S$(J): NEXT J
50   DATA f =,R =,Vt =,It =,theta =,Z =,L =,XL =,C =,Xc =,VL =,Vc =,Vr =,Ir =,
     Ic =,IL =,P =
60  FOR J=1 TO 17: READ U$(J): NEXT J
70   DATA hertz,ohms,volts,amperes,degrees,ohms,henrys,ohms,microfarads,ohms,
     volts,volts,volts,amperes,amperes,amperes,watts
80   PRINT "This program determines all the unknown values for series or
           parallel RC, RL or RCL circuits.": PRINT
90   PRINT "Press the enter key if the value of a variable is unknown.": PRINT
100   INPUT "Is the circuit a series circuit (1) or a parallel circuit (2)";Q
110  IF Q>2 THEN GOTO 100
120  INPUT "Is the circuit an RC (1), RL (2) or RCL (3) circuit";T
130  IF T>3 THEN GOTO 120
140  IF T>1 THEN D=8 ELSE D=6
150  FOR J=A TO D
160  PRINT "In ";U$(J);", "; S$(J);:INPUT I(J)
170  C(J)=I(J)
180  NEXT J
190  IF T<>2 THEN J=9:GOSUB 1650
200  IF T<>2 THEN J=10:GOSUB 1650
210  IF T>1 AND Q=1 THEN J=11:GOSUB 1650
220  IF T<>2 AND Q=1 THEN J=12:GOSUB 1650
230  IF Q=1 THEN J=13:GOSUB 1650 ELSE J=14: GOSUB 1650
240  IF T <> 2 AND Q=2 THEN J=15: GOSUB 1650
250  IF T>1 AND Q=2 THEN J=16:GOSUB 1650
260  C(5)=C(5)/57.2958
270  PRINT :PRINT "One moment please!"
280  IF C(10)=0 AND C(1)>0 AND C(9)>0 THEN C(10)=159155!/(C(1)*C(9))  'all cir
290  IF C(9)=0 AND C(1)>0 AND C(10)>0 THEN C(9)=159155!/(C(1)*C(10))
300  IF C(1)=0 AND C(9)>0 AND C(10)>0 THEN C(1)=159155!/(C(9)*C(10))
310  IF C(8)=0 THEN C(8)=6.2832*C(1)*C(7)
```

```
320 IF C(1)=0 AND C(7)>0 THEN C(1)=C(8)/(6.2832*C(7))
330 IF C(7)=0 AND C(1)>0 THEN C(7)=C(8)/(6.2832*C(1))
340 IF C(6)=0 AND C(4)>0 THEN C(6)=C(3)/C(4)
350 IF C(3)=0 THEN C(3)=C(4)*C(6)
360 IF C(4)=0 AND C(6)>0 THEN C(4)=C(3)/C(6)
370 IF Q=2 GOTO 910        'to parallel routine              start all series
380 IF C(13)=0 THEN C(13)=C(4)*C(2)
390 IF C(2)=0 AND C(4)>0 THEN C(2)=C(13)/C(4)
400 IF C(4)=0 AND C(2)>0 THEN C(4)=C(13)/C(2)
410 IF C(17)=0 THEN C(17)=C(4)*C(4)*C(2)
420 IF C(11)=0 THEN C(11)=C(4)*C(8)
430 IF C(12)=0 THEN C(12)=C(4)*C(10)
440 IF C(4)=0 AND C(8)>0 THEN C(4)=C(11)/C(8)
450 IF C(4)=0 AND C(10)>0 THEN C(4)=C(12)/C(10)
460 IF C(8)=0 AND C(4)>0 THEN C(8)=C(11)/C(4)
470 IF C(10)=0 AND C(4)>0 THEN C(10)=C(12)/C(4)
480 IF C(13)=0 AND C(5)>0 THEN C(13)=COS(C(5))*C(3)
490 IF C(2)=0 AND C(5)>0 THEN C(2)=COS(C(5))*C(6)
500 IF C(3)=0 AND C(5)>0 THEN C(3)=C(13)/COS(C(5))
510 IF C(6)=0 AND C(5)>0 THEN C(6)=C(2)/COS(C(5))
520 IF T<>1 THEN 640 'to series rl                           start series rc
530 IF C(5)=0 AND C(2)>0 THEN C(5)=ATN(C(10)/C(2))
540 IF C(5)=0 AND C(13)>0 THEN C(5)=ATN(C(12)/C(13))
550 IF C(3)=0 AND C(5)>0 THEN C(3)=C(12)/SIN(C(5))
560 IF C(6)=0 AND C(5)>0 THEN C(6)=C(10)/SIN(C(5))
570 IF C(12)=0 THEN C(12)=C(3)*SIN(C(5))
580 IF C(10)=0 THEN C(10)=C(6)*SIN(C(5))
590 IF C(13)=0 AND C(12)<C(3) AND C(12)>0 THEN C(13)=SQR(C(3)*C(3)-(C(12)*C(12))
)
600 IF C(12)=0 AND C(13)<C(3) AND C(13)>0 THEN C(12)=SQR(C(3)*C(3)-(C(13)*C(13))
)
610 IF C(10)=0 AND C(2)<C(6) AND C(2)>0 THEN C(10)=SQR(C(6)*C(6)-(C(2)*C(2)))
620 IF C(2)=0 AND C(10)<C(6) AND C(10)>0 THEN C(2)=SQR(C(6)*C(6)-(C(10)*C(10)))
630 GOTO 1430
640 IF T<>2 THEN 760 'to series rcl                          start series rl
650 IF C(5)=0 AND C(2)>0 THEN C(5)=ATN(C(8)/C(2))
660 IF C(5)=0 AND C(13)>0 THEN C(5)=ATN(C(11)/C(13))
670 IF C(3)=0 AND C(5)>0 THEN C(3)=C(11)/SIN(C(5))
680 IF C(6)=0 AND C(5)>0 THEN C(6)=C(8)/SIN(C(5))
690 IF C(12)=0 THEN C(11)=C(3)*SIN(C(5))
700 IF C(8)=0 THEN C(8)=C(6)*SIN(C(5))
710 IF C(13)=0 AND C(11)<C(3) AND C(11)>0 THEN C(13)=SQR(C(3)*C(3)-(C(11)*C(11))
)
720 IF C(11)=0 AND C(13)<C(3) AND C(13)>0 THEN C(11)=SQR(C(3)*C(3)-(C(13)*C(13))
)
730 IF C(8)=0 AND C(2)<C(6) AND C(2)>0 THEN C(8)=SQR(C(6)*C(6)-(C(2)*C(2)))
740 IF C(2)=0 AND C(8)<C(6) AND C(8)>0 THEN C(2)=SQR(C(6)*C(6)-(C(8)*C(8)))
750 GOTO 1430
760 IF X=0 AND C(10)>0 AND C(8)>0 THEN X=ABS(C(10)-C(8))    'start series rcl
770 IF VX=0 AND C(12)>0 AND C(11)>0 THEN VX=ABS(C(12)-C(11))
780 IF X=0 AND C(6)>C(2) AND C(2)>0 THEN X=SQR(C(6)*C(6)-(C(2)*C(2)))
790 IF VX=0 AND C(3)>C(13) AND C(13)>0 THEN VX=SQR(C(3)*C(3)-(C(13)*C(13)))
800 IF C(2)=0 AND C(6)>X AND X>0 THEN C(2)=SQR(C(6)*C(6)-(X*X))
810 IF C(6)=0 AND C(5)>0 THEN C(6)=X/SIN(C(5))
820 IF C(3)=0 AND C(5)>0 THEN C(3)=VX/SIN(C(5))
830 IF C(5)=0 AND C(13)>0 THEN C(5)=ATN(VX/C(13))
840 IF C(5)=0 AND C(2)>0 THEN C(5)=ATN(X/C(2))
850 IF C(8)=0 AND X>0 AND C(10)>0 THEN C(8)=ABS(X-C(10))
860 IF C(10)=0 AND X>0 AND C(8)>0 THEN C(10)=ABS(X-C(8))
870 IF C(12)=0 AND VX>0 AND C(11)>0 THEN C(12)=ABS(VX-C(11))
880 IF C(11)=0 AND VX>0 AND C(12)>0 THEN C(11)=ABS(VX-C(12))
890 IF X=0 AND C(5)>0 THEN X=SIN(C(5))*C(6)
900 GOTO 1430
910 IF C(14)=0 AND C(2)>0 THEN C(14)=C(3)/C(2)              'all parallel cir
920 IF C(2)=0 AND C(14)>0 THEN C(2)=C(3)/C(14)
930 IF C(3)=0 THEN C(3)=C(14)*C(2)
940 IF C(16)=0 AND C(8)>0 THEN C(16)=C(3)/C(8)
950 IF C(8)=0 AND C(16)>0 THEN C(8)=C(3)/C(16)
960 IF C(3)=0 THEN C(3)=C(16)*C(8)
970 IF C(15)=0 AND C(10)>0 THEN C(15)=C(3)/C(10)
980 IF C(10)=0 AND C(15)>0 THEN C(10)=C(3)/C(15)
990 IF C(3)=0 THEN C(3)=C(15)*C(10)
1000 IF C(17)=0 AND C(2)>0 THEN C(17)=C(3)*C(3)/C(2)
1010 IF C(14)=0 AND C(5)>0 THEN C(14)=COS(C(5))*C(4)
1020 IF C(4)=0 AND C(5)>0 THEN C(4)=C(14)/COS(C(5))
1030 IF C(2)=0 AND C(5)>0 THEN C(2)=C(6)/COS(C(5))
1040 IF C(6)=0 AND C(5)>0 THEN C(6)=COS(C(5))*C(2)
1050 IF T<>1 THEN 1170 'to parallel RL                       start parallel rc
1060 IF C(5)=0 AND C(10)>0 THEN C(5)=ATN(C(2)/C(10))
1070 IF C(5)=0 AND C(14)>0 THEN C(5)=ATN(C(15)/C(14))
1080 IF C(4)=0 AND C(5)>0 THEN C(4)=C(15)/SIN(C(5))
```

```
1090 IF C(15)=0 AND C(5)>0 THEN C(15)=C(4)*SIN(C(5))
1100 IF C(6)=0 THEN C(6)=C(10)*SIN(C(5))
1110 IF C(10)=0 AND C(5)>0 THEN C(10)=C(6)/SIN(C(5))
1120 IF C(14)=0 AND C(4)>C(15) AND C(15)>0 THEN C(14)=SQR(C(4)*C(4)-C(15)*C(15))
1130 IF C(15)=0 AND C(4)>C(14) AND C(14)>0 THEN C(15)=SQR(C(4)*C(4)-C(14)*C(14))
1140 IF C(2)=0 AND C(6)<C(10) AND C(6)>0 THEN C(2)=1/SQR(1/(C(6)*C(6))-1/(C(10)*C(10)))
1150 IF C(10)=0 AND C(6)<C(2) AND C(6)>0 THEN C(10)=1/SQR(1/(C(6)*C(6))-1/(C(2)*C(2)))
1160 GOTO 1430
1170 IF T<>2 THEN 1290 'to parallel RCL                    start parallel rl
1180 IF C(5)=0 AND C(8)>0 THEN C(5)=ATN(C(2)/C(8))
1190 IF C(5)=0 AND C(14)>0 THEN C(5)=ATN(C(16)/C(14))
1200 IF C(4)=0 AND C(5)>0 THEN C(4)=C(16)/SIN(C(5))
1210 IF C(16)=0 AND C(5)>0 THEN C(16)=C(4)*SIN(C(5))
1220 IF C(6)=0 THEN C(6)=C(8)*SIN(C(5))
1230 IF C(8)=0 AND C(5)>0 THEN C(8)=C(6)/SIN(C(5))
1240 IF C(14)=0 AND C(4)>C(16) AND C(16)>0 THEN C(14)=SQR(C(4)*C(4)-C(16)*C(16))
1250 IF C(16)=0 AND C(4)>C(13) AND C(13)>0 THEN C(16)=SQR(C(4)*C(4)-C(13)*C(13))
1260 IF C(2)=0 AND C(6)<C(8) AND C(6)>0 THEN C(2)=1/SQR(1/(C(6)*C(6))-1/(C(8)*C(8)))
1270 IF C(8)=0 AND C(6)<C(2) AND C(6)>0 THEN C(8)=1/SQR(1/(C(6)*C(6))-1/(C(2)*C(2)))
1280 GOTO 1430
1290 IF X=0 AND C(10)>0 AND C(8)>0 AND C(10)<>C(8) THEN X=1/ABS(1/C(10)-1/C(8))
     'par rcl
1300 IF IX=0 AND C(15)>0 AND C(16)>0 THEN IX=ABS(C(15)-C(16))
1310 IF X=0 AND C(6)<C(2) AND C(6)>0 THEN X=1/SQR(1/(C(6)*C(6))-1/(C(2)*C(2)))
1320 IF IX=0 AND C(4)>C(14) AND C(14)>0 THEN IX=SQR(C(4)*C(4)-(C(14)*C(14)))
1330 IF C(2)=0 AND C(6)<X AND C(6)>0 THEN C(2)=1/SQR(1/(C(6)*C(6))-1/(X*X))
1340 IF C(5)=0 AND C(14)>0 THEN C(5)=ATN(IX/C(14))
1350 IF C(5)=0 AND X>0 THEN C(5)=ATN(C(2)/X)
1360 IF C(6)=0 AND C(5)>0 THEN C(6)=X*SIN(C(5))
1360 IF C(6)=0 AND C(5)>0 THEN C(6)=X*SIN(C(5))
1370 IF C(4)=0 AND C(5)>0 THEN C(4)=IX/SIN(C(5))
1380 IF X=0 AND C(5)>0 THEN X=C(6)/SIN(C(5))
1390 IF C(8)=0 AND X>0 AND X<C(10) THEN C(8)=1/ABS(1/X-1/C(10))
1400 IF C(10)=0 AND X>0 AND X<C(8) THEN C(10)=1/ABS(1/X-1/C(8))
1410 IF C(15)=0 AND IX>0 AND C(16)>0 THEN C(15)=ABS(IX-C(16))
1420 IF C(16)=0 AND IX>0 AND C(15)>0 THEN C(16)=ABS(IX-C(15))
1430 S=0
1440 FOR J=1 TO 17
1450 IF ABS(C(J)-E(J))>.00001 THEN S=1
1460 E(J)=C(J)
1470 NEXT J
1480 IF S=1 GOTO 280
1490 C(5)=C(5)*57.2958
1500 CLS: PRINT "The entered values were:"
1510 FOR J=1 TO 16
1520 IF I(J)<>0 THEN PRINT S$(J);I(J);U$(J)
1530 NEXT J
1540 PRINT: PRINT "The calculated values are:"
1550 IF C(8)=C(10) AND X=0 THEN C(6)=C(2): C(2)=C(6)
1560 FOR J=1 TO 17
1570 IF I(J)=0 AND C(J) <>0 THEN PRINT S$(J);C(J);U$(J)
1580 NEXT J
1590 IF C(6)=C(2) AND C(6)>0 THEN PRINT "theta = 0 degrees
1600 IF Q=2 AND C(5)>0 AND ((C(15)>C(16)) OR (C(8)>C(10) AND C(10)>0)) THEN PRINT "It leads Vt"
1610 IF Q=2 AND C(5)>0 AND ((C(15)<C(16)) OR (C(8)<C(10) AND C(8)>0)) THEN PRINT "It lags Vt"
1620 IF Q=1 AND C(5)>0 AND ((C(12)>C(11)) OR (C(10)>C(8) AND C(8)>0)) THEN PRINT "Vt lags It"
1630 IF Q=1 AND C(5)>0 AND ((C(12)<C(11)) OR (C(10)<C(8) AND C(10)>0)) THEN PRINT "Vt leads It"
1640 GOTO 1670
1650 PRINT "In ";U$(J);", ";S$(J);:INPUT I(J)
1660 C(J)=I(J):RETURN
1670 CLEAR:PRINT:LOCATE 24,1
1680 PRINT "DO ANOTHER? (Y/N)";
1690 X$ = INKEY$:IF X$="" THEN 1690
1700 IF X$ <> "Y" AND X$ <> "y" AND X$ <> "N" AND X$ <>"n" THEN PRINT "WRONG KEY":GOTO 1690
1710 IF X$ = "Y" OR X$ ="y" THEN 10 ELSE END
```

SAMPLE RUNS FOR *RCL* CIRCUITS

```
                        RCL CIRCUITS

This program determines all the unknown values for series or
parallel RC, RL or RCL circuits.

Press the enter key if the value of a variable is unknown.

Is the circuit a series circuit (1) or a parallel circuit (2)? 1
Is the circuit an RC (1), RL (2) or RCL (3) circuit? 3
In hertz, f =? 1200
In ohms, R =? 47
In volts, Vt =? 10
In amperes, It =?
In degrees, theta =?
In ohms, Z =?
In henrys, L =? .15
In ohms, XL =?
In microfarads, C =? .22
In ohms, Xc =?
In volts, VL =?
In volts, Vc =?
In volts, Vr =?




The entered values were:
f = 1200 hertz
R = 47 ohms
Vt = 10 volts
L = .15 henrys
C = .22 microfarads

The calculated values are:
It = 1.886067E-02 amperes
theta = 84.91436 degrees
Z = 530.204 ohms
XL = 1130.976 ohms
Xc = 602.8599 ohms
VL = 21.33096 volts
Vc = 11.37034 volts
Vr = .8864513 volts
P = 1.671906E-02 watts
Vt leads It



DO ANOTHER? (Y/N)


                        RCL CIRCUITS

This program determines all the unknown values for series or
parallel RC, RL or RCL circuits.

Press the enter key if the value of a variable is unknown.

Is the circuit a series circuit (1) or a parallel circuit (2)? 2
Is the circuit an RC (1), RL (2) or RCL (3) circuit? 2
In hertz, f =? 500
In ohms, R =? 100
In volts, Vt =? 30
In amperes, It =?
In degrees, theta =?
In ohms, Z =?
In henrys, L =?
In ohms, XL =? 120
In amperes, Ir =?
In amperes, IL =?
```

```
The entered values were:
f = 500 hertz
R = 100 ohms
Vt = 30 volts
XL = 120 ohms

The calculated values are:
It = .3905125 amperes
theta = 39.80559 degrees
Z = 76.82213 ohms
L = .0381971 henrys
Ir = .3 amperes
IL = .25 amperes
P = 9 watts
It lags Vt

DO ANOTHER? (Y/N)
```

```
                    RCL CIRCUITS

This program determines all the unknown values for series or
parallel RC, RL or RCL circuits.

Press the enter key if the value of a variable is unknown.

Is the circuit a series circuit (1) or a parallel circuit (2)? 1
Is the circuit an RC (1), RL (2) or RCL (3) circuit? 3
In hertz, f =? 25000
In ohms, R =? 2200
In volts, Vt =? 10
In amperes, It =?
In degrees, theta =? 20
In ohms, Z =?
In henrys, L =?
In ohms, Xl =? 1400
In microfarads, C =?
In ohms, Xc =?
In volts, VL =?
In volts, Vc =?
In volts, Vr =?


The entered values were:
f = 25000 hertz
R = 2200 ohms
Vt = 10 volts
theta = 20 degrees
XL = 1400 ohms

The calculated values are:
It = 4.271331E-03 amperes
Z = 2341.191 ohms
L = 8.912656E-03 henrys
C = 1.062333E-02 microfarads
Xc = 599.266 ohms
VL = 5.979862 volts
Vc = 2.559663 volts
Vr = 9.396927 volts
P = 4.013738E-02 watts
Vt leads It

DO ANOTHER? (Y/N)
```

PROGRAM 18-1 RESONANCE

This program computes L, C, or f for either series or parallel circuits when two of the three variables are specified. It also computes Q, BW, f_{up} and f_{lo} when the appropriate resistance is specified. When no resistances are specified, the program skips the Q, BW, f_{up}, and f_{lo} calculations.

The upper and lower limits of the BW are calculated using the approximation formulas $f_{up} = f_r + 0.5\text{BW}$ and $f_{lo} = f_r - 0.5\text{BW}$. The program is accurate for all series resonant circuits. However, for parallel circuits, the standard formula $f_r = 1/(2\pi\sqrt{LC})$ is used. This means that for an inductor Q of 10, an error of about 0.5 percent is introduced. With a Q of 4, the error will be about 3 percent.

```
10 CLS: SCREEN 0: WIDTH 80: CLEAR
20 PRINT TAB(40) "RESONANCE": PRINT
30 A$(1) = "series": A$(2) = "parallel"
40 PRINT "This program computes f, L or C for series and parallel circuits."
50 PRINT "If a requested resistance value is unknown, press the enter key."
60 PRINT
70 INPUT "Series circuit (1) or parallel circuit (2)"; T
80 IF T<1 OR T>2 THEN PRINT "WRONG KEY":GOTO 70
90 INPUT "Is the unknown quantity the f (1), L (2) or C (3)"; U
100 IF U<1 OR U>3 THEN PRINT "WRONG KEY":GOTO 90
110 IF U<>2 THEN INPUT "The inductance (in henries) is"; L
120 INPUT "The resistance (in ohms) of the inductor is"; RL
130 PRINT "The value (in ohms) of the physical resistor in "A$(T)" with LC is";
140 INPUT R
150 IF U>1 THEN INPUT "The frequency (in hertz) is"; F
160 IF U<3 THEN INPUT "The capacitance (in farads) is"; C
170 IF F = 0 THEN F = 1/(SQR(L*C)*6.283)
180 IF L = 0 THEN L = .02533/(F*F*C)
190 IF C = 0 THEN C = .02533/(F*F*L)
200 XL = 6.283*F*L
210 IF T = 1 AND (RL>0 OR R>0) THEN Q = XL/(R+RL)
220 IF T = 2 AND R>0 AND RL>0 THEN Q = 1/(RL/XL+XL/R)
230 IF T = 2 AND R = 0 AND RL>0 THEN Q = XL/RL
240 IF Q>0 THEN BW = F/Q
250 FL = F-.5*BW
260 FU = F+.5*BW
270 PRINT: PRINT
280 IF U = 1 THEN PRINT "The frequency is" F "hertz."
290 IF U = 2 THEN PRINT "The inductance is" L "henries."
300 IF U = 3 THEN PRINT "The capacitance is" C "farads"
310 IF Q = 0 THEN 370
320 IF Q<2 THEN PRINT: PRINT "This is not a practical circuit (Q<2).": GOTO 370
330 PRINT "The quality is" Q
340 PRINT "The bandwidth is" BW "hertz."
350 PRINT "The upper frequency of the bandwidth is" FU "hertz."
360 PRINT "The lower frequency of the bandwidth is" FL "hertz."
370 CLEAR:PRINT:LOCATE 24,1
380 PRINT "DO ANOTHER? (Y/N)";
390 X$ = INKEY$:IF X$="" THEN 390
400 IF X$ <> "Y" AND X$ <> "y" AND X$ <> "N" AND X$ <>"n" THEN PRINT "WRONG KEY"
:GOTO 390
410 IF X$ = "Y" OR X$ ="y" THEN 10 ELSE END
```

SAMPLE RUNS FOR RESONANCE

```
                                      RESONANCE

This program computes f, L or C for series and parallel circuits.
If a requested resistance value is unknown, press the enter key.

Series circuit (1) or parallel circuit (2)? 1
Is the unknown quantity the f (1), L (2) or C (3)? 1
The inductance (in henries) is? .001
The resistance (in ohms) of the inductor is?
The value (in ohms) of the physical resistor in series with LC is?
The capacitance (in farads) is? 220e-12


The frequency is 339329.5 hertz.


DO ANOTHER? (Y/N)
```

```
                                      RESONANCE

This program computes f, L or C for series and parallel circuits.
If a requested resistance value is unknown, press the enter key.

Series circuit (1) or parallel circuit (2)? 2
```

```
Is the unknown quantity the f (1), L (2) or C (3)? 3
The inductance (in henries) is? .001
The resistance (in ohms) of the inductor is? 15
The value (in ohms) of the physical resistor in parallel with LC is?
The frequency (in hertz) is? 256000


The capacitance is 3.865051E-10 farads
The quality is 107.2299
The bandwidth is 2387.394 hertz.
The upper frequency of the bandwidth is 257193.7 hertz.
The lower frequency of the bandwidth is 254806.3 hertz.


DO ANOTHER? (Y/N)
```

PROGRAM 19-1 COMPLEX NUMBERS

This program provides seven options: (1) polar-to-rectangular conversion, (2) rectangular-to-polar conversion, (3) addition of two phasors, (4) subtraction of two phasors, (5) multiplication of two phasors, (6) division of two phasors, and (7) termination of program.

As soon as the selected option is executed, the program returns to the option menu and awaits the selection of another option. This process continues until option 7 is selected. For each option, the data entered by the operator and the answer provided by the program are printed on the screen and saved until the screen is filled. Seven operations can be performed and printed before the screen is filled (see the last sample run).

The first two sample runs are divided into screens. Whenever "options": appears it is the start of a new screen. For example, in the first sample run the first screen ends when the "190.2" is entered. When the enter key is pressed to enter "190.2," the program deletes part of the material printed on the left half of the screen and adds the answer section on the right half of the screen.

In the second sample, the screen changes six times as the operator inputs the data for subtracting two phasors. In the last sample run only the final screen is shown after adding, multiplying, converting, subtracting, dividing, and again adding and multiplying. The percentage signs (%) in the last answer merely indicate that the answer is too large to fit the format specified by the program. Note that polar and rectangular forms can be mixed for any of the arithmetical operations and that the form of the answer can be specified.

```
10 CLS: SCREEN 0: WIDTH 80: CLEAR
20 DIM U(2),V(2)
30 PRINT "Options:"
40 PRINT "  1. Polar to rectangular"
50 PRINT "  2. Rectangular to polar"
60 PRINT "  3. Add two phasors"
70 PRINT "  4. Subtract two phasors"
80 PRINT "  5. Multiply two phasors"
90 PRINT "  6. Divide two phasors"
100 PRINT "  7. End program"
110 PI=3.141593
120 FOR I=11 TO 24
130    LOCATE I,1: PRINT SPACE$(39);
140 NEXT I
150 LOCATE 11,1: INPUT "Option? ",C
160 LOCATE 1,25: PRINT SPACE$(15)
170 IF C=7 THEN CLS:END
180 IF C<>1 THEN 250
190    LOCATE 19,1: INPUT "Magnitude? ",R
200    LOCATE 20,1: INPUT "Angle? ",A
210    LOCATE N+1,40: PRINT "Phasor 1 :";: GOSUB 730
220    A=PI*A/180
230    X=R*COS(A): Y=R*SIN(A)
240    LOCATE N+1,43: PRINT "Answer:";: GOSUB 740: GOTO 120
250 IF C<>2 THEN 350
260    LOCATE 19,1: INPUT "Real component: ",X
270    LOCATE 20,1: INPUT "Imaginary component: ",Y
280    LOCATE N+1,40: PRINT "Phasor 1 :";: GOSUB 740
290    R=SQR(X*X+Y*Y)
300    A=ATN(Y/X)
```

```
310    A=180*A/PI
320    IF X<0 THEN A=A+180
330    IF A>180 THEN A=A-360
340    LOCATE N+1,43: PRINT "Answer:";: GOSUB 730: GOTO 120
350 IF C<>3 AND C<>4 AND C<>5 AND C<>6 THEN 120
360 LOCATE 13,1: PRINT "Do you want answer to be in"
370 LOCATE 14,7: PRINT "(1) Rectangular coordinates, or"
380 LOCATE 15,7: PRINT "(2) Polar coordinates?";
390 LOCATE 16,8: PRINT SPACE$(10);: LOCATE 16,8: INPUT P
400 IF P<>1 AND P<>2 THEN 380
410 FOR I=1 TO 2
420    LOCATE 13,1: PRINT SPACE$(39);
430    LOCATE 13,1: PRINT "Will phasor number";I;"be in"
440    LOCATE 16,7: PRINT SPACE$(20);: LOCATE 16,8: INPUT Q
450    IF Q<>1 AND Q<>2 THEN 440
460    IF Q=2 THEN 540
470       LOCATE 19,1: PRINT "Real component:"; SPACE$(15);
480       LOCATE 19,17: INPUT "",U(I)
490       LOCATE 20,1: PRINT "Imaginary component:"; SPACE$(10);
500       LOCATE 20,22: INPUT "",V(I)
510       X=U(I): Y=V(I): LOCATE N+1,40: PRINT "Phasor";I;":";: GOSUB 740
520       LOCATE 19,1: PRINT SPACE$(35);: LOCATE 20,1: PRINT SPACE$(35);
530       GOTO 600
540       LOCATE 19,1: PRINT "Magnitude:"; SPACE$(20);: LOCATE 19,12: INPUT "",R
550       LOCATE 20,1: PRINT "Angle:"; SPACE$(24);: LOCATE 20,8: INPUT "",A
560       LOCATE N+1,40: PRINT "Phasor";I;":";: GOSUB 730
570       LOCATE 19,1: PRINT SPACE$(35);: LOCATE 20,1: PRINT SPACE$(35);
580       A=PI*A/180
590       U(I)=R*COS(A): V(I)=R*SIN(A)
600 NEXT I
610 ON C-2 GOTO 630, 640, 650, 660
620 GOTO 120
630 X=U(1)+U(2): Y=V(1)+V(2): IF P=2 THEN 290 ELSE 240
640 X=U(1)-U(2): Y=V(1)-V(2): IF P=2 THEN 290 ELSE 240
650 X=U(1)*U(2)-V(1)*V(2): Y=U(1)*V(2)+U(2)*V(1): IF P=2 THEN 290 ELSE 240
660 IF U(2)<>0 OR V(2)<>0 THEN 700
670    BEEP: LOCATE 22,5: PRINT "Cannot divide by zero"
680    PRINT "Press any key to continue"
690    X$=INKEY$: IF X$="" THEN 690 ELSE LOCATE 23,1 :PRINT SPACE$(25): GOTO 120
700 X=(U(1)*U(2)+V(1)*V(2))/(U(2)*U(2)+V(2)*V(2))
710 Y=(U(2)*V(1)-U(1)*V(2))/(U(2)*U(2)+V(2)*V(2))
720 IF P=2 THEN 290 ELSE 240
730 PRINT USING "######.###    angle ####.###";R;A: N=N+1: RETURN
740 IF Y>=0 THEN PRINT USING "######.### _+ _j######.###";X;Y
750 IF Y<0 THEN Y=-Y: PRINT USING "######.### _- _j######.###";X;Y: Y=-Y
760 N=N+1: RETURN
```

SAMPLE RUNS FOR COMPLEX NUMBERS

```
Options:
   1. Polar to rectangular
   2. Rectangular to polar
   3. Add two phasors
   4. Subtract two phasors
   5. Multiply two phasors
   6. Divide two phasors
   7. End program


Option? 2


Real component: 256.9
Imaginary component: 190.2


Options:                             Phasor 1 :    256.900 + j    190.200
   1. Polar to rectangular              Answer:    319.646   angle   36.515
   2. Rectangular to polar
   3. Add two phasors
   4. Subtract two phasors
   5. Multiply two phasors
   6. Divide two phasors
   7. End program


Option?
```

```
Options:
    1. Polar to rectangular
    2. Rectangular to polar
    3. Add two phasors
    4. Subtract two phasors
    5. Multiply two phasors
    6. Divide two phasors
    7. End program

Option? 4

Do you want answer to be in
        (1) Rectangular coordinates, or
        (2) Polar coordinates?
         ? 2


Options:
    1. Polar to rectangular
    2. Rectangular to polar
    3. Add two phasors
    4. Subtract two phasors
    5. Multiply two phasors
    6. Divide two phasors
    7. End program

Option? 4

Will phasor number 1 be in
        (1) Rectangular coordinates, or
        (2) Polar coordinates?
         ? 1


Options:
    1. Polar to rectangular
    2. Rectangular to polar
    3. Add two phasors
    4. Subtract two phasors
    5. Multiply two phasors
    6. Divide two phasors
    7. End program

Option? 4

Will phasor number 1 be in
        (1) Rectangular coordinates, or
        (2) Polar coordinates?
         ? 1

Real component: 258.2
Imaginary component: 394.7

Options:                                      Phasor 1 :    258.200 + j    394.700
    1. Polar to rectangular
    2. Rectangular to polar
    3. Add two phasors
    4. Subtract two phasors
    5. Multiply two phasors
    6. Divide two phasors
    7. End program

Option? 4

Will phasor number 2 be in
        (1) Rectangular coordinates, or
        (2) Polar coordinates?
         ? 2
```

Options:
   1. Polar to rectangular
   2. Rectangular to polar
   3. Add two phasors
   4. Subtract two phasors
   5. Multiply two phasors
   6. Divide two phasors
   7. End program

Phasor 1 :   258.200 + j   394.700

Option? 4

Will phasor number 2 be in
     (1) Rectangular coordinates, or
     (2) Polar coordinates?
      ? 2

Magnitude: 189.5
Angle: 37.6

Options:
   1. Polar to rectangular
   2. Rectangular to polar
   3. Add two phasors
   4. Subtract two phasors
   5. Multiply two phasors
   6. Divide two phasors
   7. End program

Phasor 1 :   258.200 + j   394.700
Phasor 2 :   189.500   angle   37.600
 Answer: 299.268   angle   68.833

Option?

Options:
   1. Polar to rectangular
   2. Rectangular to polar
   3. Add two phasors
   4. Subtract two phasors
   5. Multiply two phasors
   6. Divide two phasors
   7. End program

Phasor 1 :   234.000   angle   56.000
Phasor 2 :   245.000 - j   289.000
 Answer: 387.673   angle -14.186
Phasor 1 :   786.000 + j   45.000
Phasor 2 :   23.000   angle 145.000
 Answer: 18107.600   angle 148.277
Phasor 1 : 18107.600   angle 148.277
 Answer: -15402.330 + j   9521.210
Phasor 1 :   134.000 - j   56.000
Phasor 2 :   65.000   angle -32.000
 Answer: 81.769   angle -15.284
Phasor 1 :   81.769   angle -15.284
Phasor 2 :   387.673   angle -14.186
 Answer: 0.211 - j   0.004
Phasor 1 :   34.000   angle   67.000
Phasor 2 :   34.000 + j   89.000
 Answer: 47.285 + j   120.297
Phasor 1 : 76855.000 + j 35656.000
Phasor 2 : 45677.000 - j 45675.000
 Answer:%5139094000.000 - j%1881693000.

Option?

PROGRAM 19-2   AC NETWORK ANALYSIS

This program uses nodal analysis to determine the currents and voltages in a complex ac circuit. The nodes being used should be consecutively numbered starting with 0 for the reference node.

The sample run is for a circuit containing a parallel R and L [R(3,0) and L(3,0)] connected in series with a R and C [R(1,2) and C(2.3)]. The circuit is powered by a 100 kHz, 30 V /0° ac source. The circuit diagram, with the modes labeled, is shown at the beginning of the sample run.

```
10 CLS: SCREEN 0: WIDTH 80: CLEAR
20 DIM B(10,40),A(20,40)
30 PRINT TAB(30) "AC NETWORK ANALYSIS"
40 PRINT TAB(25) "(Node zero is the reference node)":PRINT
50 INPUT"ENTER THE NUMBER OF NODES";N1
60 NC=N1-1
70 INPUT "ENTER THE FREQUENCY IN HERTZ";F
80 INPUT"ENTER THE NUMBER OF RESISTORS";R
90 IF R = 0 THEN 190
100 FOR K=1 TO R
110 PRINT"ENTER THE VALUE FOR RESISTOR";K;"IN OHMS";
120 INPUT X
130 PRINT"ENTER THE TWO NODES CONNECTED TO RESISTOR";K;
140 INPUT L,M
150 R(L,M)=X
160 A(L,M)=(-1/X)+A(L,M)+A(M,L)
170 A(M,L)=0
180 NEXT K
190 PRINT
200 PI = 3.141592654#
210 INPUT "ENTER THE NUMBER OF SOURCES";S
220 FOR K=1 TO S
230 PRINT"ENTER THE VALUE OF SOURCE";K;"IN VOLTS";
240 INPUT X
250 PRINT "ENTER ITS PHASE ANGLE IN DEGREES";:INPUT T
260 INPUT "ENTER ITS POSITIVE NODE";M
270 INPUT "ENTER ITS NEGATIVE NODE";L
280 I(L,M)=10000!*X*COS(PI*T/180)
290 I1(L,M)=10000!*X*SIN(PI*T/180)
300 A(L,M)=A(L,M)-10000! + A(M,L)
310 A(M,L)=0
320 NEXT K
330 GOSUB 1580
340 P = 1
350 GOSUB 1710
360 GOSUB 1670
370 FOR J = 1 TO NC
380 K = 2*J-1
390 FOR I = 1 TO NC
400 B(I,K)=A(I,J)
410 NEXT I: NEXT J
420 FOR I = 0 TO NC:FOR J = 0 TO NC
430 A(I,J) = 0
440 NEXT J: NEXT I
450 W = 2*PI*F
460 PRINT
470 INPUT "ENTER THE NUMBER OF INDUCTORS";L1
480 IF L1 = 0 THEN 590
490 FOR K = 1 TO L1
500 PRINT"ENTER THE VALUE FOR INDUCTOR";K;"IN HENRIES";
510 INPUT X
520 XL =W*X
530 PRINT"ENTER THE TWO NODES CONNECTED TO INDUCTOR";K;
540 INPUT L,M
550 L(L,M)=XL
560 A(L,M)=(1/XL)+A(L,M)+A(M,L)
570 A(M,L)=0
580 NEXT K
590 PRINT
600 PRINT"ENTER THE NUMBER OF CAPACITORS";
610 INPUT C
620 IF C = 0 THEN 730
630 FOR K=1 TO C
640 PRINT"ENTER THE VALUE FOR CAPACITOR";K;"IN FARADS";
650 INPUT X
660 XC=X*W
670 PRINT"ENTER THE TWO NODES CONNECTED TO CAPACITOR";K;
680 INPUT L,M
690 C(L,M)=1/XC
700 A(L,M)=(-XC)+A(L,M)+A(M,L)
710 A(M,L)=0
720 NEXT K
730 GOSUB 1580
```

```
740 FOR J= 1 TO NC
750 K= 2*J-1
760 FOR I= 1 TO NC
770 B(I,K+1) =A(I,J)
780 NEXT I: NEXT J
790 FOR I = 1 TO NC:FOR J = 1 TO NC
800 A(I,J)=0
810 NEXT J: NEXT I
820 PRINT
830 N=2*NC
840 MA=N-1
850 MB=N+1
860 MC=N
870 MD=N+2
880 IK = -1
890 FOR I = 1 TO MA STEP 2
900 IK = IK+1
910 FOR J=1 TO MB STEP 2
920 LJ = I-IK
930 A(I,J) = B(LJ,J)
940 A(I,J+1) = -B(LJ,J+1)
950 NEXT J:NEXT I
960 FOR I = 2 TO MC STEP 2
970 FOR J = 1 TO MB STEP 2
980 LJ = INT(I/2)
990 A(I,J) = B(LJ,J+1)
1000 A(I,J+1) = B(LJ,J)
1010 NEXT J:NEXT I
1020 PRINT "COMPUTING THE MAGNITUDE AND PHASE ANGLE AT EACH NODE"
1030 EP = .000001
1040 M=N+1
1050 KK=0
1060 JJ=0
1070 FOR K = 1 TO N
1080 JJ = KK+1
1090 LL=JJ
1100 KK = KK+1
1110 IF ABS(A(JJ,KK))-EP>0 THEN 1140
1120 JJ = JJ+1
1130 GOTO 1110
1140 IF LL-JJ = 0 THEN 1190
1150 FOR MM = 1 TO M
1160 AT=A(LL,MM)
1170 A(LL,MM) = A(JJ,MM)
1180 A(JJ,MM) = AT:NEXT MM
1190 FOR LJ = 1 TO M
1200 J = M+1-LJ
1210 A(K,J) = A(K,J)/A(K,K):NEXT LJ
1220 FOR I = 1 TO N
1230 FOR LJ = 1 TO M
1240 J = M+1-LJ
1250 IF (I-K) = 0 THEN 1270
1260 A(I,J) = A(I,J)-A(I,K)*A(K,J)
1270 NEXT LJ:NEXT I:NEXT K
1280 PRINT
1290 FOR I = 1 TO NC
1300 K = 2*I-1
1310 X = SQR(A(K,M)^2+A(K+1,M)^2)
1320 IF A(K,M)=0 THEN T=90 ELSE T=ATN(A(K+1,M)/A(K,M))*180/PI
1330 IF A(K,M)<0 THEN T =T+180
1340 PRINT"VOLTAGE";I;"=";:PRINT USING "##.##^^^^";X;:PRINT " AT AN ANGLE OF ";:
PRINT USING "####.##";T;:PRINT " DEGREES"
1350 NEXT I
1360 PRINT:PRINT "PRESS ANY KEY TO CONTINUE"
1370 G$ = INKEY$:IF G$ = "" THEN 1370
1380 A$ = "R"
1390 FOR I = 0 TO NC
1400 FOR J = 0 TO NC
1410 I(I,J)=R(I,J)
1420 NEXT J: NEXT I
1430 P=0:PRINT
1440 GOSUB 1820
1450 A$="L"
1460 FOR I = 0 TO NC:FOR J = 0 TO NC
1470 I(I,J)=L(I,J)
1480 NEXT J: NEXT I
1490 P = -90
1500 GOSUB 1820
1510 A$ = "C"
1520 FOR I = 0 TO NC:FOR J = 0 TO NC
1530 I(I,J)=C(I,J)
```

```
1540 NEXT J: NEXT I
1550 P = 90
1560 GOSUB 1820
1570 GOTO 1940
1580 FOR I = 1 TO NC:FOR J = 0 TO NC
1590 IF I = J THEN 1610
1600 A(I,I)=A(I,I)-A(I,J)-A(J,I)
1610 NEXT J: NEXT I
1620 FOR I = 0 TO NC:FOR J = 0 TO NC
1630 IF I = J THEN 1650
1640 IF A(I,J)=0 THEN A(I,J)=A(J,I) ELSE A(J,I)=A(I,J)
1650 NEXT J: NEXT I
1660 RETURN
1670 FOR I = 0 TO NC
1680 FOR J = 0 TO NC
1690 I(I,J)=I1(I,J)
1700 NEXT J: NEXT I
1710 FOR J = 0 TO NC:FOR K = 0 TO NC
1720 IF J=K THEN 1740
1730 IF I(K,J)=0 THEN I(K,J)=-I(J,K) ELSE I(J,K)=-I(K,J)
1740 NEXT K: NEXT J
1750 N = 2*NC
1760 FOR J = 1 TO NC:FOR K = 0 TO NC
1770 IF J=K THEN 1790
1780 B(J,N+P)=B(J,N+P)+I(K,J)
1790 NEXT K: NEXT J
1800 P = P+1
1810 RETURN
1820 FOR I = 0 TO NC:FOR J = 0 TO NC
1830 IF I(I,J)=0 THEN 1920
1840 IF I = 0 THEN ER = -A(2*J-1,M):EI = -A(2*J,M):GOTO 1880
1850 IF J = 0 THEN ER = A(2*I-1,M):EI = A(2*I,M):GOTO 1880
1860 ER = A((2*I)-1,M)-A((2*J)-1,M)
1870 EI = A(2*I,M)-A(2*J,M)
1880 C=SQR(ER^2+EI^2)/I(I,J)
1890 IF ER=0 THEN T=P+90 ELSE T=ATN(EI/ER)*180/PI + P
1900 IF ER<0 THEN T=T+180
1910 PRINT"THE CURRENT THRU ";A$;"(";I;",";J;") IS";:PRINT USING "##.##^^^^";C;:
PRINT " WITH AN ANGLE OF ";:PRINT USING "####.##";T;:PRINT " DEGREES"
1920 NEXT J: NEXT I
1930 RETURN
1940 CLEAR:PRINT:LOCATE 24,1
1950 PRINT "DO ANOTHER? (Y/N)"
1960 X$ = INKEY$:IF X$="" THEN 1960
1970 IF X$ <> "Y" AND X$ <> "y" AND X$ <> "N" AND X$ <>"n" THEN PRINT "WRONG KEY ":GOTO 1960
1980 IF X$ = "Y" OR X$ ="y" THEN 10 ELSE END
```

SAMPLE RUN FOR AC NETWORK ANALYSIS

The following is a schematic diagram for the sample run of Program 19-2.

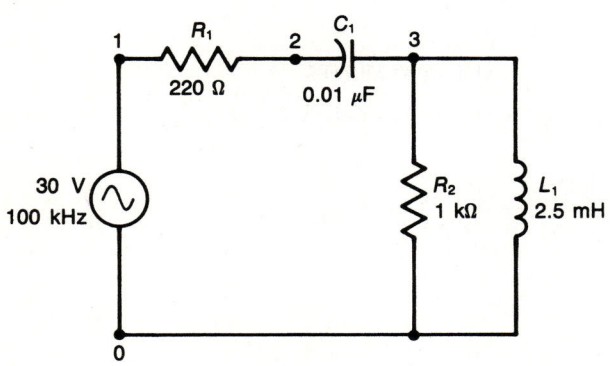

AC NETWORK ANALYSIS
(Node zero is the reference node)

```
ENTER THE NUMBER OF NODES? 4
ENTER THE FREQUENCY IN HERTZ? 100000
ENTER THE NUMBER OF RESISTORS? 2
ENTER THE VALUE FOR RESISTOR 1 IN OHMS? 220
ENTER THE TWO NODES CONNECTED TO RESISTOR 1 ? 1,2
```

```
ENTER THE VALUE FOR RESISTOR 2 IN OHMS? 1000
ENTER THE TWO NODES CONNECTED TO RESISTOR 2 ? 3,0

ENTER THE NUMBER OF SOURCES? 1
ENTER THE VALUE OF SOURCE 1 IN VOLTS? 30
ENTER ITS PHASE ANGLE IN DEGREES? 0
ENTER ITS POSITIVE NODE? 1
ENTER ITS NEGATIVE NODE? 0

ENTER THE NUMBER OF INDUCTORS? 1
ENTER THE VALUE FOR INDUCTOR 1 IN HENRIES? .0025
ENTER THE TWO NODES CONNECTED TO INDUCTOR 1 ? 3,0

ENTER THE NUMBER OF CAPACITORS? 1
ENTER THE VALUE FOR CAPACITOR 1 IN FARADS? .00000001
ENTER THE TWO NODES CONNECTED TO CAPACITOR 1 ? 2,3

COMPUTING THE MAGNITUDE AND PHASE ANGLE AT EACH NODE

VOLTAGE 1 = 3.00E+01 AT AN ANGLE OF    0.00 DEGREES
VOLTAGE 2 = 2.36E+01 AT AN ANGLE OF    4.93 DEGREES
VOLTAGE 3 = 2.59E+01 AT AN ANGLE OF   14.97 DEGREES

PRESS ANY KEY TO CONTINUE

THE CURRENT THRU R( 1 , 2 ) IS 3.07E-02 WITH AN ANGLE OF  -17.51 DEGREES
THE CURRENT THRU R( 3 , 0 ) IS 2.59E-02 WITH AN ANGLE OF   14.97 DEGREES
THE CURRENT THRU L( 3 , 0 ) IS 1.65E-02 WITH AN ANGLE OF  -75.03 DEGREES
THE CURRENT THRU C( 2 , 3 ) IS 3.07E-02 WITH AN ANGLE OF  342.49 DEGREES

DO ANOTHER? (Y/N)
```

PROGRAM 20-1   TRANSFORMER ANALYSIS

This program will determine the primary current, power factor, efficiency, load current, and load voltage for a power transformer. The required operator entries are the data from the open-circuit test ($V_{P_{OC}}$, $I_{P_{OC}}$, and $P_{P_{OC}}$) and the closed-circuit test ($V_{P_{SC}}$, $I_{P_{SC}}$, and $P_{P_{SC}}$), the no-load secondary voltage, and the load impedance (both magnitude and angle).

The analysis is based on standard equivalent circuits. It takes into account both the core loss (and its effective resistance R_C) and the primary self-inductance X_M.

Voltages, currents, powers, and impedance are entered in base units. The phase angle of the load must be entered in degrees.

```
10  CLS : SCREEN 0: WIDTH 80: CLEAR
20    PRINT TAB(30); "TRANSFORMER ANALYSIS": PRINT
30    PRINT "This program determines PF, IP, n, VL and IL for a specified load."
40    PRINT "The input requires the data from a standard transformer test."
50    PRINT
60    INPUT "Open-circuit test voltage "; VOC
70    INPUT "Open-circuit test current "; IOC
80    INPUT "Open-circuit test power "; POC
90    INPUT "Short-circuit test voltage "; VSC
100    INPUT "Short-circuit test current "; ISC
110    INPUT "Short-circuit test power "; PSC
120   INPUT "No-load secondary voltage "; VSOC
130   INPUT "Load impedance magnitude "; ZLM
140   INPUT "Load impedance angle "; ZLA
150   PRINT
160   PI = 3.141593
170   RE = PSC / (ISC * ISC)
180   A = VOC / VSOC
190   XE = SQR(VSC / ISC * (VSC / ISC) - (RE * RE))
200   ZLMR = ZLM * A * A
210   ZLAR = ZLA * PI / 180
220   CO = COS(ZLAR)
230   SI = SIN(ZLAR)
240   RT = RE + CO * ZLMR
250   XT = XE + SI * ZLMR
260   Z = SQR(RT * RT + XT * XT)
270   IZ = VOC / Z
280   IF IZ = 0 THEN VS = VSOC: IW = 0 ELSE VS = IZ * ZLMR / A: IW = VS / ZLM
```

```
290 EFF = IZ * IZ * CO * ZLMR * 100 / (IZ * IZ * RT + POC)
300 RC = VOC * VOC / POC
310 ZOC = VOC / IOC
320 XM = SQR(ZOC * ZOC * RC * RC / (RC * RC - ZOC * ZOC))
330 XTP = (XT * XT + RT * RT) / XT
340 RTP = (RT * RT + XT * XT) / RT
350 RTT = RTP * RC / (RTP + RC)
360 XTT = 1 / (1 / XM + 1 / XTP)
370 PZT = SQR(RTT * RTT + XTT * XTT)
380 ZT = RTT * XTT / PZT
390 THZT = ATN(1 / XTT / (1 / RTT)): PF = COS(THZT)
400 IP = VOC / ZT
410 PRINT USING "The power factor is #.###"; PF
420 PRINT USING "The primary current is ###.##_ A"; IP
430 PRINT USING "The % of efficiency is ##.#_ %"; EFF
440 PRINT USING "The load voltage is ####.##_ V"; VS
450 PRINT USING "The load current is ####.##_ A"; IW
460 CLEAR : PRINT : LOCATE 24, 1
470 PRINT "DO ANOTHER? (Y/N)";
480 X$ = INKEY$: IF X$ = "" THEN 480
490 IF X$ <> "Y" AND X$ <> "y" AND X$ <> "N" AND X$ <> "n" THEN PRINT "WRONG KEY
": GOTO 480
500 IF X$ = "Y" OR X$ = "y" THEN 10 ELSE END
```

SAMPLE RUN FOR TRANSFORMER ANALYSIS

```
                        TRANSFORMER ANALYSIS

This program determines PF, IP, n, VL and IL for a specified load.
The input requires the data from a standard transformer test.

Open-circuit test voltage ? 120
Open-circuit test current ? .08
Open-circuit test power ? 4
Short-circuit test voltage ? 10
Short-circuit test current ? .6
Short-circuit test power ? 5.8
No-load secondary voltage ? 18
Load impedance magnitude ? 15
Load impedance angle ? 20

The power factor is 0.832
The primary current is    0.24 A
The % of efficiency is 81.1 %
The load voltage is   17.56 V
The load current is    1.17 A


DO ANOTHER? (Y/N)
```

PROGRAM 21-1 MULTIPLIERS AND SHUNTS

The resistances needed to convert a meter movement to a multirange voltmeter and/or ammeter are calculated by this program. The user must input the meter movement ratings and the range values. The output provides the sensitivity rating for the voltage ranges as well as the multiplier resistances.

The limit of six ranges each for voltage and current can be changed by changing the 6 in lines 130 and 230. If the limit is to be 10 or more ranges, then a DIM VR(NVR) and DIM IR(NIR) must be inserted as lines 135 and 235, respectively.

```
10 CLS: SCREEN 0: WIDTH 80: CLEAR
20 PRINT TAB(30) "MULTIPLIERS AND SHUNTS":PRINT
30 PRINT "This program determines multiplier and shunt resistances for"
40 PRINT "a maximum of six voltage and six current ranges."
50 PRINT
60 PRINT "ENTER DATA IN BASE UNITS"
70 PRINT
80 INPUT "Ifs (full-scale current of meter movement) ";IFS
90 INPUT "Ri (internal resistance of meter movement) ";RI
100 INPUT "Number of voltage ranges (0 to 6) ";NVR
```

```
110 VFS = IFS * RI
120 IF NVR = 0 THEN 210
130 IF NVR > 6 OR NVR < 0 THEN 100
140 FOR I = 1 TO NVR
150 PRINT "Voltage range";I;
160 INPUT VR(I)
170 IF VR(I) > VFS THEN 200
180 PRINT "Voltage ranges must be >";VFS
190 GOTO 150
200 NEXT I
210 INPUT "Number of current ranges (0 to 6) ";NIR
220 IF NIR = 0 THEN 320
230 IF NIR > 6 OR NIR < 0 THEN 210
240 FOR I = 1 TO NIR
250 PRINT "Current range";I;
260 INPUT IR(I)
270 IF IR(I) > IFS THEN 300
280 PRINT "Current ranges must be >";IFS
290 GOTO 250
300 NEXT I
310 PRINT
320 IF NVR = 0 THEN 390
330 PRINT "Voltage range"; TAB(30) "Multiplier Resistance"
340 FOR I = 1 TO NVR
350 PRINT VR(I)"volts"; TAB(30) (VR(I)-VFS)/IFS"ohms"
360 NEXT I
370 PRINT "Sensitivity = "1/IFS;"ohms/volt"
380 PRINT
390 IF NIR = 0 THEN 440
400 PRINT "Current Range" TAB(30) "Shunt Resistance"
410 FOR I = 1 TO NIR
420 PRINT IR(I)"amperes"; TAB(30) VFS/(IR(I)-IFS)"ohms"
430 NEXT I :PRINT
440 CLEAR:PRINT:LOCATE 24,1
450 PRINT "DO ANOTHER? (Y/N)";
460 X$ = INKEY$:IF X$="" THEN 460
470 IF X$ <> "Y" AND X$ <> "y" AND X$ <> "N" AND X$ <>"n" THEN PRINT "WRONG KEY"
:GOTO 460
480 IF X$ = "Y" OR X$ ="y" THEN 10 ELSE END
```

SAMPLE RUNS FOR MULTIPLIERS AND SHUNTS

```
                        MULTIPLIERS AND SHUNTS

This program determines multiplier and shunt resistances for
a maximum of six voltage and six current ranges.

ENTER DATA IN BASE UNITS

Ifs (full-scale current of meter movement) ? .001
Ri (internal resistance of meter movement) ? 75
Number of voltage ranges (0 to 6) ? 0
Number of current ranges (0 to 6) ? 3
Current range 1 ? .01
Current range 2 ? .1
Current range 3 ? 1

Current Range                  Shunt Resistance
 .01 amperes                    8.333334 ohms
 .1 amperes                     .7575758 ohms
 1 amperes                      7.507508E-02 ohms


DO ANOTHER? (Y/N)



                        MULTIPLIERS AND SHUNTS

This program determines multiplier and shunt resistances for
a maximum of six voltage and six current ranges.

ENTER DATA IN BASE UNITS

Ifs (full-scale current of meter movement) ? .00005
Ri (internal resistance of meter movement) ? 1500
Number of voltage ranges (0 to 6) ? 6
```

```
Voltage range 1 ? 2.5
Voltage range 2 ? 10
Voltage range 3 ? 25
Voltage range 4 ? 50
Voltage range 5 ? 100
Voltage range 6 ? 500
Number of current ranges (0 to 6) ? 6
Current range 1 ? .0001
Current range 2 ? .001
Current range 3 ? .01
Current range 4 ? .1
Current range 5 ? 1
Current range 6 ? 10


Voltage range            Multiplier Resistance
 2.5 volts                 48500 ohms
 10 volts                  198500 ohms
 25 volts                  498500 ohms
 50 volts                  998500 ohms
 100 volts                 1998500 ohms
 500 volts                 9998500 ohms
Sensitivity =   20000 ohms/volt

Current Range            Shunt Resistance
 .0001 amperes             1500 ohms
 .001 amperes              78.94736 ohms
 .01 amperes               7.537689 ohms
 .1 amperes                .7503751 ohms
 1 amperes                 7.500375E-02 ohms
 10 amperes                7.500037E-03 ohms

DO ANOTHER? (Y/N)
```

PROGRAM 23-1 POWER SUPPLY COMPONENTS

Selecting an appropriate diode, capacitor, and transformer to provide a desired output voltage, output current, and ripple factor requires considerable manipulation of data to obtain reasonably accurate results. This program uses some empirically derived formulas which compensate for such factors as (1) the change in the ripple waveform as the ripple factor changes and (2) decreases in transformer output voltage as the load current increases. This latter factor can be quite significant when the peak recurrent current is many times greater than the current rating of the transformer.

The transformer voltages calculated by the program are close estimates of the rms voltage when the transformer is providing the current required by the load. Often this current is considerably less than the rated current of the transformer. In this case, the transformer's actual output voltage may be considerably higher than its rated value. Therefore, the dc output voltage from the power supply may be slightly greater than predicted by the program unless you use a transformer that has an actual value close to that specified by the program.

```
10 CLS : SCREEN 0: WIDTH 80: CLEAR
20 PRINT TAB(30); "POWER SUPPLY COMPONENTS": PRINT
30 PRINT "This program will determine the value of the components needed to prov
   ide a    specified dc voltage and dc current at a specified ripple factor when
   the ac   supply is 60 hertz.": PRINT
40 INPUT "Will the power supply be half-wave (1),  full-wave c-t (2) or bridge (
   3)"; T
50 IF T < 1 OR T > 3 THEN 40
60 INPUT "Desired dc output voltage"; V
70 IF V = 0 OR V < 0 OR V > 10000 THEN PRINT "VOLTAGE MUST BE >0 AND <10,000.":
   GOTO 60
80 INPUT "Load current (in amperes)"; I
90 IF I > 10 THEN PRINT "Load current can not be greater than 10 A": GOTO 80
100 INPUT "Ripple factor"; R
110 IF R < 0 OR R > .5 THEN PRINT "Ripple factor must be >0 and <0.5": GOTO 100
120 PRINT
130 DIM A(12): A(0) = .1: A(1) = .25: A(2) = .5: A(3) = 1: A(4) = 2: A(5) = 4: A
```

```
    (6) = 6: A(7) = 8: A(8) = 10: A(9) = 12: A(10) = 16: A(11) = 20: A(12) = 24
140 IF T = 1 THEN F = 60 ELSE F = 120
150 IF T = 3 THEN VD = 1.4 ELSE VD = .7
160 K = (SIN(R + .9)) - .79
170 C = (1000000! / (3.5 * R * (V / I) * F)) * (1 - 1.5 * K)
180 H = (LOG(1 / (2 * R))) / 2.3
190 E = V + 1.75 * V * R + VD
200 FOR J = 0 TO 9: IT = A(J)
210 IF IT > I GOTO 230
220 NEXT J
230 FOR L = J TO 12: IT = A(L)
240 G = (LOG(100 * I / IT)) / 23
250 CF = G * H * .6
260 EPT = (E * (1 + CF))
270 ET = .707 * EPT
280 Y = SQR(R * 100)
290 IP = (200 / (8.399999 * Y)) * I
300 WV% = EPT: PIV% = 2 * EPT
310 ID = 1.11 * I
320 IF T = 2 THEN ET = 1.414 * EPT: IT = IT / 2: ID = ID / 2: IP = IP / 2
330 IF T = 3 THEN ID = ID / 2: IP = IP / 2
340 PT = I * V
350 IF PT > IT * ET THEN J = J + 1: GOTO 230
360 M = M + 1
370 IF M < 2 THEN PRINT "For a safty factor, increase all diode and capacitor ra
tings about 15% and then select the next higher available values.": PRINT
380 IF M < 2 THEN PRINT "Capacitor ratings: "; C; "microfarads,"; WV%; "DCWV"
390 IF M < 2 THEN PRINT "Diode ratings:  PIV="; PIV%; "V, If="; ID; "A, and Ifpr
="; IP; "A"
400 IF M < 2 THEN PRINT "Transformer ratings -- any of the following:"
410 IF T = 2 THEN PRINT , ET; "Vc-t, "; IT; "A" ELSE PRINT , ET; "V, ", IT; "A"
420 IF CF < .02 THEN PRINT , " Same voltage as above, and any current greater th
an above.": PRINT : GOTO 440
430 NEXT L: PRINT
440 CLEAR : LOCATE 24, 1
450 PRINT "DO ANOTHER? (Y/N)";
460 X$ = INKEY$: IF X$ = "" THEN 460
470 IF X$ <> "Y" AND X$ <> "y" AND X$ <> "N" AND X$ <> "n" THEN PRINT "WRONG KEY
": GOTO 460
480 IF X$ = "Y" OR X$ = "y" THEN 10 ELSE END
```

SAMPLE RUNS FOR POWER SUPPLY COMPONENTS

POWER SUPPLY COMPONENTS

This program will determine the value of the components needed to provide a specified dc voltage and dc current at a specified ripple factor when the ac supply is 60 hertz.

Will the power supply be half-wave (1), full-wave c-t (2) or bridge (3)? 2
Desired dc output voltage? 18

Load current (in amperes)? 6
Ripple factor? .36

For a safty factor, increase all diode and capacitor ratings about 15% and then select the next higher available values.

Capacitor ratings: 1668.572 microfarads, 31 DCWV
Diode ratings: PIV= 61 V, If= 3.33 A, and Ifpr= 11.90476 A
Transformer ratings -- any of the following:
            43.15987 Vc-t, 4 A
            Same voltage as above, and any current greater than above.

DO ANOTHER? (Y/N)

PROGRAM 23-2 ZENER REGULATOR COMPONENTS

This program will determine (1) the resistance and power rating of the series resistor and (2) the power rating of the zener diode needed to construct a zener-regulated circuit. The program will work for zener voltages between 1.8 and

200 V and for zener power ratings up to 10 W. The operator must supply the zener voltage V_z, the maximum load current to be drawn from the regulated supply I_{LM}, the maximum input voltage to the regulator V_{dcm}, and the minimum input voltage to the regulator V_{min}. In determining V_{min}, remember that V_{min} is the minimum instantaneous voltage; i.e., the value present when the ripple is at its maximum negative swing. The minimum load current for the regulator is assumed to be zero. Then the zener will not be overloaded even as the load on the regulator circuit develops an open circuit.

The power ratings specified for the zener diode and the resistor have been derated at least 50 percent to allow for a safety factor. This should be sufficient derating unless extremely inadequate heat-sinking and air circulation are encountered.

```
10 CLS: SCREEN 0: WIDTH 80: CLEAR
20 PRINT TAB(25) "ZENER REGULATOR CONPONENTS":PRINT
30 PRINT "This program determines the zener diode power rating and the resistanc
e and     power rating of the series resistor." :PRINT :PRINT
40 INPUT "Voltage rating of the zener diode"; VZ
50 IF VZ <1.8 OR VZ >200 THEN PRINT "Voltage must be >1.7 and < 201 volts":GOTO
    40
60 INPUT "Maximum regulator load current"; IL
70 INPUT "Maximum input voltage to the regulator"; VMA
80 IF VMA=<VZ THEN PRINT "Maximum input must be > the diode voltage":GOTO 70
90 INPUT "Minimum input voltage to the regulator"; VMI
100 IF VMI=>VMA THEN PRINT "Minimum input must be < maximum input":GOTO 90
110 IF VMI=<VZ THEN PRINT "Minimum input must be > the diode voltage":GOTO 90
120 PRINT: PRINT
130 A(0)=.25:A(1)=.4:A(2)=.5:A(3)=1:A(4)=5:A(5)=10
140 B(0)=.5:B(1)=1:B(2)=2:B(3)=3:B(4)=5:B(5)=8:B(6)=10:B(7)=12:B(8)=25:B(9)=50
150 IDM=40*IL/((40*VMI-40*VZ)/(VMA-VZ)-1) : IF IDM=<0 GOTO 320
160 PD=IDM*VZ*2
170 IF PD>10 GOTO 320
180 FOR J=0 TO 5: PDA=A(J)
190 IF PDA>.95*PD GOTO 210
200 NEXT J
210 IZMA=PDA/VZ:IZMI=IZMA*.025
220 RMI=(VMA-VZ)/(.8*IZMA)
230 RMA=(VMI-VZ)/(IZMI+IL)
240 PR=((VMA-VZ)^2/RMA)*2
250 IF RR>50 GOTO 340
260 FOR J=0 TO 9: PRA=B(J)
270 IF PRA>.95 * PR GOTO 290
280 NEXT J
290 PRINT "Use a" PDA "W,"VZ"V zener."
300 PRINT "Use a"PRA"W resistor with a resistance between"RMA"ohms and"RMI"ohms
but as close to" RMA"ohms as practical."
310 PRINT:GOTO 350
320 PRINT "Requires a zener power rating greater than 10 W.  Use a different typ
e of      regulator circuit."
330 PRINT:GOTO 350
340 PRINT "Requires a resistor power rating greater than 50W.  Use a different t
ype of regulator circuit.":PRINT
350 CLEAR:PRINT:LOCATE 24,1
360 PRINT "DO ANOTHER?  (Y/N)" ;
370 X$ = INKEY$:IF X$="" THEN 370
380 IF X$ <> "Y" AND X$ <> "y" AND X$ <> "N" AND X$ <>"n" THEN PRINT "WRONG KEY"
:GOTO 370
390 IF X$ = "Y" OR X$ ="y" THEN 10 ELSE END
```

SAMPLE RUNS FOR ZENER REGULATOR COMPONENTS

```
                    ZENER REGULATOR CONPONENTS
This program determines the zener diode power rating and the resistance and
power rating of the series resistor.


Voltage rating of the zener diode? 14
Maximum regulator load current? .08
Maximum input voltage to the regulator? 22
Minimum input voltage to the regulator? 18
```

Use a 5 W, 14 V zener.
Use a 3 W resistor with a resistance between 44.97992 ohms and 28
ohms but as close to 44.97992 ohms as practical.

DO ANOTHER? (Y/N)

                    ZENER REGULATOR CONPONENTS

This program determines the zener diode power rating and the resistance and
power rating of the series resistor.

Voltage rating of the zener diode? 14
Maximum regulator load current? 2
Maximum input voltage to the regulator? 30
Minimum input voltage to the regulator? 24

Requires a zener power rating greater than 10 W. Use a different type of
regulator circuit.

DO ANOTHER? (Y/N)

PROGRAM 24-1   CE AMPLIFIER ANALYSIS

Beta (β) is used in this program to calculate the input resistance, but in most circuits of this type it has only minimum affect on r_i so its value is not critical. If minimum β for the transistor is known, use that value; otherwise, use the typical value. If β is not known, use 50, which is a value exceeded by most small-signal transistors. In this program R_{B_1} is the resistor connected between base and V_{CC} and R_{B_2} is connected from base to common. On the screen, R_{B_1} will appear as RB1 because BASIC doesn't support subscripts.

This program requires that the user furnish values for β, R_L, R_C, R_E bypassed, R_E unbypassed, R_{B_1}, R_{B_2}, and V_{CC}. If the amplifier is not loaded, enter 1,000,000 for R_L. Enter 0 for R_E unbypassed or R_E bypassed if one or the other is not present in the circuit you wish to analyze. Enter all resistances in ohms and enter V_{CC} in volts.

The output values for V_B, V_C, and V_E are not preceded by $+$ or $-$ signs; however, they are the values from these leads to common or ground. It is left to the user to insert the appropriate polarity sign for these voltages.

```
10 CLS: SCREEN 0: WIDTH 80: CLEAR
20 PRINT TAB(25) "CE AMPLIFIER ANALYSIS":PRINT
30 PRINT "This program determines AV, AI, AP, ri, VB, VC and VE for a beta-indep
endent,";
40 PRINT "single-supply, CE amplifier."
50 PRINT
60 INPUT "VCC (collector supply voltage)"; VCC
70 IF VCC<0 THEN PRINT "USE ABSOLUTE VALUE OF VCC":GOTO 60
80 INPUT "RB1 (base-to-VCC resistor)"; RB1
90 INPUT "RB2 (base-to-common resistor)"; RB2
100 INPUT "Unbypassed RE"; REU
110 INPUT "Bypassed RE"; REB
120 INPUT "RC (collector resistor)"; RC
130 INPUT "RL (load resistance)"; RL
140 INPUT "Beta"; B
150 PRINT
160 VB=VCC*RB2/(RB2+RB1)
170 VE = VB-.7
180 IF VE=<0 THEN PRINT "Transistor is cutoff.":GOTO 350
190 IC=VE/(REU+REB) :ICM=1000*IC
200 IF VE+(IC*RC)=>VCC THEN PRINT "Transistor is saturated.":GOTO 350
210 RE=.026/IC
220 VC=VCC-(IC*RC)
230 AV=(RC*RL/(RC+RL))/(RE+REU)
240 RI%=1/((1/RB1)+(1/RB2)+(1/((B+1)*(RE+REU))))
250 AI=(AV/RL)/(1/RI%)
260 AP=AI*AV
```

```
270 PRINT USING "The voltage gain is ###.##";AV
280 PRINT USING "The current gain is ###.##";AI
290 PRINT USING "The power gain is #####.##";AP
300 PRINT "The input resistance is" RI% "ohms"
310 PRINT USING "The collector voltage is ###.## _V"; VC
320 PRINT USING "The emitter voltage is ###.## _V"; VE
330 PRINT USING "The base voltage is ###.## _V"; VB
340 PRINT USING "The collector current is ####.### _m_A"; ICM
350 CLEAR:PRINT:LOCATE 24,1
360 PRINT "DO ANOTHER? (Y/N)" ;
370 X$ = INKEY$:IF X$="" THEN 370
380 IF X$ <> "Y" AND X$ <> "y" AND X$ <> "N" AND X$ <>"n" THEN PRINT "WRONG KEY"
:GOTO 370
390 IF X$ = "Y" OR X$ ="y" THEN 10 ELSE END
```

SAMPLE RUNS FOR CE AMPLIFIER ANALYSIS

```
                    CE AMPLIFIER ANALYSIS

This program determines AV, AI, AP, ri, VB, VC and VE for a beta-independent,
single-supply, CE amplifier.

VCC (collector supply voltage)? 25
RB1 (base-to-VCC resistor)? 270000
RB2 (base-to-common resistor)? 33000
Unbypassed RE? 470
Bypassed RE? 1800
RC (collector resistor)? 10000
RL (load resistance)? 20000
Beta? 120

The voltage gain is   13.36
The current gain is   13.21
The power gain is    176.38
The input resistance is 19777 ohms
The collector voltage is  16.09 V
The emitter voltage is    2.02 V
The base voltage is    2.72 V
The collector current is    0.891 mA

DO ANOTHER? (Y/N)


                    CE AMPLIFIER ANALYSIS

This program determines AV, AI, AP, ri, VB, VC and VE for a beta-independent,
single-supply, CE amplifier.

VCC (collector supply voltage)? 9
RB1 (base-to-VCC resistor)? 100000
RB2 (base-to-common resistor)? 10000
Unbypassed RE? 200
Bypassed RE? 0
RC (collector resistor)? 4700
RL (load resistance)? 5100
Beta? 100

The voltage gain is   10.02
The current gain is   13.05
The power gain is    130.85
The input resistance is 6641 ohms
The collector voltage is   6.22 V
The emitter voltage is    0.12 V
The base voltage is   0.82 V
The collector current is    0.591 mA

DO ANOTHER? (Y/N)
```

PROGRAM 25-1 DIFFERENTIAL AMPLIFIER ANALYSIS

The analysis performed by this program is valid for unloaded differential amplifiers operating at low frequencies (audio range). It is assumed that the dc resistance from base to common will be less than 10 kΩ.

The program requires the user to furnish absolute values, in base units, for V_{CC}, V_{EE}, R_C, R_E, and β. Because BASIC does not allow for subscripts, V_{CC} will appear as VCC, etc. Also, B will be used in place of β. The output from the program provides values for A_D, A_{CM}, C_{MRR}, Z_i, V_C, and I_C. Because a differential amplifier may use either PNP or NPN transistors, no polarity sign will precede the values of V_C in the output. Z_i is dependent upon β, so if you input the minimum value for β, the output will provide the minimum value of Z_i also.

```
10 CLS: SCREEN 0: WIDTH 80: CLEAR
20 PRINT TAB(25) "DIFFERENTIAL AMPLIFIER ANALYSIS":PRINT
30 PRINT "This program determines AD, ACM, CMRR, Zi, VC, and IC for a ";
40 PRINT "differential amplifier."
50 PRINT
60 INPUT "VCC (collector supply voltage)"; VCC
70 IF VCC<0 THEN PRINT "USE ABSOLUTE VALUE OF VCC":GOTO 60
80 INPUT "VEE (emitter supply voltage)"; VEE
90 IF VEE<0 THEN PRINT "USE ABSOLUTE VALUE OF VEE":GOTO 80
100 INPUT "RC (collector resistor)"; RC
110 INPUT "RE (emitter resistor)"; RE
120 INPUT "Beta"; B
130 PRINT
140 IC=(VEE-.7)/(2*RE) : ICM=IC*1000
150 IF IC*RC=>VCC THEN PRINT "Transistors are saturated.":GOTO 280
160 VC=VCC-(IC*RC)
170 RED=.026/IC
180 AD=RC/(2*RED) : AD%=AD
190 ACM=RC/(2*(RE+RED))
200 CMRR%=AD/ACM
210 ZI=B*2*RED/1000
220 PRINT "The differential voltage gain (AD) is" AD%
230 PRINT USING "The common-mode voltage gain (ACM) is #.###"; ACM
240 PRINT "The common-mode rejection ratio (CMRR) is" CMRR%
250 PRINT "The input impedance (Zi) is" ZI"k ohms"
260 PRINT USING "The collector voltage (VC) is ###.## _V"; VC
270 PRINT USING "The collector current is ###.### _m_A"; ICM
280 CLEAR:PRINT:LOCATE 24,1
290 PRINT "DO ANOTHER? (Y/N)" ;
300 X$ = INKEY$:IF X$="" THEN 300
310 IF X$ <> "Y" AND X$ <> "y" AND X$ <> "N" AND X$ <>"n" THEN PRINT "WRONG KEY"
:GOTO 300
320 IF X$ = "Y" OR X$ ="y" THEN 10 ELSE END
```

SAMPLE RUNS FOR DIFFERENTIAL AMPLIFIER ANALYSIS

```
                    DIFFERENTIAL AMPLIFIER ANALYSIS

This program determines AD, ACM, CMRR, Zi, VC, and IC for a
differential amplifier.

VCC (collector supply voltage)? 14
VEE (emitter supply voltage)? 14
RC (collector resistor)? 82000
RE (emitter resistor)? 120000
Beta? 160

The differential voltage gain (AD) is 87
The common-mode voltage gain (ACM) is 0.340
The common-mode rejection ratio (CMRR) is 257
The input impedance (Zi) is 150.1353 k ohms
The collector voltage (VC) is   9.46 V
The collector current is    0.055 mA


DO ANOTHER? (Y/N)



                    DIFFERENTIAL AMPLIFIER ANALYSIS

This program determines AD, ACM, CMRR, Zi, VC, and IC for a
differential amplifier.
```

```
VCC (collector supply voltage)? 12
VEE (emitter supply voltage)? 12
RC (collector resistor)? 150000
RE (emitter resistor)? 68000
Beta? 120

Transistors are saturated.

DO ANOTHER? (Y/N)
```

PROGRAM 25-2 NONINVERTING AMPLIFIER ANALYSIS

This program analyzes a noninverting amplifier and determines the voltage gain, input impedance, bandwidth, and peak-to-peak output voltage. Peak-to-peak output can be limited by either the slew rate of the op amp or the value of V_{CC} and V_{EE}. For this program, it is assumed that the output of the op amp can swing to within 2 V of V_{CC} or V_{EE}.

Other conditions required for the output of this program to be valid are: (1) dual supplies with $V_{EE} = V_{CC}$, (2) sine-wave input signal, and (3) 20 db per decade roll-off.

The user of this program must provide the following information about the amplifier being analyzed: (1) supply voltage, (2) R of the feedback resistor, (3) R of the input resistor, (4) highest frequency the amplifier must amplify, (5) unity-gain frequency of the op amp, (6) the input impedance of the op amp, and (7) the slew rate of the op amp. The last three items can be obtained from the specification sheet for the op amp used in the noninverting amplifier.

```
10  CLS: SCREEN 0: WIDTH 80: CLEAR
20  PRINT TAB(25) "NONINVERTING AMPLIFIER ANALYSIS":PRINT
30  PRINT "This program determines AV, Zi, BW and undistorted Vp-p output ";
40  PRINT "for a noninverting amplifier when /VCC/ = /VEE/."
50  PRINT
60  INPUT "VCC (collector supply voltage in volts)"; VCC
70  IF VCC<3 THEN PRINT "Absolute value of VCC must be => 3.":GOTO 60
80  INPUT "Rf (feedback resistor in kilohms)"; RF
90  INPUT "R1 (input resistor in kilohms)"; R1
100 INPUT "Fu (unity-gain frequency of op amp in MHz)"; FU
110 INPUT "Fo (highest operating frequency in kHz)"; FO
120 INPUT "Zi(op) (Zi of the op amp in megohms)"; ZIOP
130 INPUT "SR (slew rate in V/microsecond)"; SR
140 PRINT
150 AV=RF/R1+1
160 ZI=FU*1000/FO*ZIOP/AV
170 BW=FU*1000/AV
180 VPPO=SR*1000/(3.2*FO)
190 IF VPPO>VCC-2 THEN VPPO=VCC-2
200 IF FO>BW THEN PRINT "AV too high for operating frequency. Reduce Rf." :
    GOTO 250
210 PRINT USING "The voltage gain (AV) is ####.##"; AV
220 PRINT USING "The input impedance (Zi) is ###.## _m_e_g_o_h_m_s"; ZI
230 PRINT USING "The bandwidth (BW) is ####.## _k_H_z"; BW
240 PRINT USING "The p-p output voltage (Vp-p out) is ##.## _V"; VPPO
250 CLEAR:PRINT:LOCATE 24,1
260 PRINT "DO ANOTHER? (Y/N)" ;
270 X$ = INKEY$:IF X$="" THEN 270
280 IF X$ <> "Y" AND X$ <> "y" AND X$ <> "N" AND X$ <>"n" THEN PRINT "WRONG KEY"
    :GOTO 270
290 IF X$ = "Y" OR X$ ="y" THEN 10 ELSE END
```

SAMPLE RUNS FOR NONINVERTING AMPLIFIER ANALYSIS

```
                    NONINVERTING AMPLIFIER ANALYSIS

This program determines AV, Zi, BW and undistorted Vp-p output
for a noninverting amplifier when /VCC/ = /VEE/.
```

```
VCC (collector supply voltage in volts)? 15
Rf (feedback resistor in kilohms)? 10
R1 (input resistor in kilohms)? 1.2
Fu (unity-gain frequency of op amp in MHz)? 2
Fo (highest operating frequency in kHz)? 20
Zi(op) (Zi of the op amp in megohms)? 1
SR (slew rate in V/microsecond)? 1

The voltage gain (AV) is     9.33
The input impedance (Zi) is   10.71 megohms
The bandwidth (BW) is  214.29 kHz
The p-p output voltage (Vp-p out) is 13.00 V


DO ANOTHER? (Y/N)


                    NONINVERTING AMPLIFIER ANALYSIS

This program determines AV, Zi, BW and undistorted Vp-p output
for a noninverting amplifier when /VCC/ = /VEE/.

VCC (collector supply voltage in volts)? 15
Rf (feedback resistor in kilohms)? 100
R1 (input resistor in kilohms)? 1
Fu (unity-gain frequency of op amp in MHz)? .75
Fo (highest operating frequency in kHz)? 30
Zi(op) (Zi of the op amp in megohms)? 2
SR (slew rate in V/microsecond)? 5

AV too high for operating frequency. Reduce Rf.

DO ANOTHER? (Y/N)
```

PROGRAM 26-1 DECIMAL TO BINARY CONVERSION

This program converts any positive decimal number to its binary equivalent. The decimal number may include a decimal point.

The binary equivalent number may have up to 20 places to the right of the binary point. If you desire to limit the answer to fewer places, change the -20 in line 90.

```
10 CLS: SCREEN 0: WIDTH 80: CLEAR
20 PRINT TAB(25) "DECIMAL TO BINARY CONVERSION":PRINT
30 PRINT "This program determines the binary equivalent of a positive decimal nu
mber."
40 PRINT: PRINT
50 INPUT "The decimal number is ",X
60 IF X<=0 THEN 50
70 Y=INT(LOG(X)/LOG(2))
80 IF Y<-1 THEN Y=-1
90 FOR I=Y+1 TO -20 STEP -1
100 IF X>0 THEN 130
110 IF I>=0 THEN A$=A$+STRING$(I+1,"0")
120 GOTO 160
130 IF I=-1 THEN A$=A$+"."
140 IF X>=2^I THEN X=X-2^I: A$=A$+"1"   ELSE IF LEN(A$)>0 THEN A$=A$+"0"
150 NEXT I
160 PRINT "The binary equivalent is "A$
170 CLEAR:PRINT:LOCATE 24,1
180 PRINT "DO ANOTHER? (Y/N)" ;
190 X$ = INKEY$:IF X$="" THEN 190
200 IF X$ <> "Y" AND X$ <> "y" AND X$ <> "N" AND X$ <>"n" THEN PRINT "WRONG KEY"
:GOTO 190
210 IF X$ = "Y" OR X$ ="y" THEN 10 ELSE END
```

SAMPLE RUNS FOR DECIMAL TO BINARY CONVERSION

```
                    DECIMAL TO BINARY CONVERSION

This program determines the binary equivalent of a positive decimal number.


The decimal number is 268
The binary equivalent is 100001100


DO ANOTHER? (Y/N)

                    DECIMAL TO BINARY CONVERSION

This program determines the binary equivalent of a positive decimal number.


The decimal number is 5768.638
The binary equivalent is 1011010001000.10100011011


DO ANOTHER? (Y/N)
```

PROGRAM 26-2 BINARY TO DECIMAL CONVERSION

The user of this program is asked to input a binary number. The binary number may include a binary point. If any digits other than 1 and 0 are entered, the program rejects the number and requests another entry.

The output from the program is the decimal equivalent of the binary number. The resolution of the output is limited only by the limitation of the computer.

```
10  CLS: SCREEN 0: WIDTH 80: CLEAR
20  PRINT TAB(25) "BINARY TO DECIMAL CONVERSION":PRINT
30  PRINT "This program determines the decimal equivalent of a binary number."
40  PRINT: PRINT
50  INPUT "The binary number is ",Y$
60  IF VAL(Y$)<=0 THEN 50
70  Y=INSTR(Y$,".")-2
80  IF Y=-2 THEN Y=LEN(Y$)-1 ELSE Y$=LEFT$(Y$,Y+1)+RIGHT$(Y$,LEN(Y$)-Y-2)
90  FOR I=1 TO LEN(Y$)
100 IF MID$(Y$,I,1)="1" THEN A=A+2^(Y-I+1) ELSE IF MID$(Y$,I,1)<>"0" THEN PRINT
    "Illegal entry": A=0: GOTO 50
110 NEXT I
120 PRINT "The decimal equivalent is"A
130 CLEAR:PRINT:LOCATE 24,1
140 PRINT "DO ANOTHER? (Y/N)" ;
150 X$ = INKEY$:IF X$="" THEN 150
160 IF X$ <> "Y" AND X$ <> "y" AND X$ <> "N" AND X$ <>"n" THEN PRINT "WRONG KEY"
    :GOTO 150
170 IF X$ = "Y" OR X$ ="y" THEN 10 ELSE END
```

SAMPLE RUNS FOR BINARY TO DECIMAL CONVERSION

```
                    BINARY TO DECIMAL CONVERSION

This program determines the decimal equivalent of a binary number.


The binary number is 1101.1101
The decimal equivalent is 13.8125


DO ANOTHER? (Y/N)
```

BINARY TO DECIMAL CONVERSION

This program determines the decimal equivalent of a binary number.

```
The binary number is 110001102.011101
Illegal entry
The binary number is 110001101.011101
The decimal equivalent is 397.4531

DO ANOTHER? (Y/N)
```

PROGRAM 26-3 BASE N TO BASE N CONVERSION

This is a general-purpose radix conversion program that will convert a number in any base to any other base you choose. Its versatility is illustrated in the sample run.

```
10 CLS: SCREEN 0: WIDTH 80: CLEAR
20 PRINT TAB(20) "BASE N TO BASE N CONVERSION":PRINT
30 PRINT"THIS PROGRAM WILL CONVERT INTEGER OR REAL NUMBERS UP TO BASE 16"
40 INPUT"WHAT IS YOUR NUMBER";B$
50 INPUT"WHAT IS ITS BASE";B
60 IF B > 16 THEN 480
70 IF B < 2 THEN 480
80 INPUT"WHAT IS THE NEW BASE";B2
90 IF B2 > 16 THEN 480
100 IF B2 < 2 THEN 480
110 Y$="."
120 L=LEN(B$)
130 P=0
140 FOR I = 1 TO L
150 P=P+1
160 Z$=MID$(B$,I,1)
170 IF Z$ = Y$ THEN 200
180 NEXT
190 P=0
200 X=P
210 E=L
220 FOR Z = 1 TO L
230 F$=MID$(B$,Z,1)
240 F=ASC(F$)
250 IF F = 46 THEN 440
260 IF F < 48 THEN 480
270 IF F > 70 THEN 480
280 IF F >= 65 THEN 310
290 IF F > 57 THEN 480
300 F=F-48:GOTO 320
310 F=F-55
320 IF F >= B THEN 480
330 GOTO 340
340 IF P = 0 THEN 360
350 GOTO 370
360 V=B^(E-1):E=E-1
370 IF P = 1 THEN 390
380 GOTO 400
390 V=B^(T-1):T=T-1
400 IF P>1 THEN 420
410 GOTO 430
420 V=B^(X-2):X=X-1
430 A=F*V:S=S+A
440 NEXT
450 IF B=10 THEN 490
460 PRINT:PRINT"YOUR NUMBER IN DECIMAL IS";S
470 GOTO 490
480 PRINT "ILLEGAL ENTRY":T=0:S=0:GOTO 40
490 IF B2=10 THEN 790
500 S1=INT(S)
510 Q=S1/B2
520 R=S1-(INT(Q)*B2)
530 IF INT(Q)=0 THEN 610
540 IF R>9 THEN 560
550 R$=STR$(R):GOTO 570
560 R$=CHR$(R+55)
570 S$=R$+S$
```

```
580 Q1=INT(Q)/B2
590 R=INT(Q)-(INT(Q1)*B2)
600 Q=Q1:GOTO 530
610 IF R > 9 THEN 630
620 GOTO 640
630 R$=CHR$(R+55):GOTO 650
640 R$=STR$(R)
650 S$=R$+S$
660 D=S-S1
670 IF P = 0 THEN 780
680 IF D = 0 THEN 780
690 S$=S$+Y$
700 FOR J = 1 TO 10
710 D=D*B2:M=INT(D):D=D-M
720 IF M > 9 THEN 740
730 M$=STR$(M):GOTO 750
740 M$=CHR$(M+55)
750 S$=S$+M$
760 IF D = 0 THEN 780
770 NEXT
780 PRINT:PRINT"YOUR NUMBER IN BASE";B2;"IS ";S$
790 CLEAR:PRINT:LOCATE 24,1
800 PRINT "DO ANOTHER? (Y/N)" ;
810 X$ = INKEY$:IF X$="" THEN 810
820 IF X$ <> "Y" AND X$ <> "y" AND X$ <> "N" AND X$ <>"n" THEN PRINT "WRONG KEY"
:GOTO 810
830 IF X$ = "Y" OR X$ ="y" THEN 10 ELSE END
```

SAMPLE RUN FOR BASE *N* TO BASE *N* CONVERSION

```
                    BASE N TO BASE N CONVERSION

THIS PROGRAM WILL CONVERT INTEGER OR REAL NUMBERS UP TO BASE 16
WHAT IS YOUR NUMBER? 56
WHAT IS ITS BASE? 10
WHAT IS THE NEW BASE? 2

YOUR NUMBER IN BASE 2 IS  1 1 1 0 0 0


DO ANOTHER? (Y/N)


                    BASE N TO BASE N CONVERSION

THIS PROGRAM WILL CONVERT INTEGER OR REAL NUMBERS UP TO BASE 16
WHAT IS YOUR NUMBER? 3FD
WHAT IS ITS BASE? 16
WHAT IS THE NEW BASE? 10

YOUR NUMBER IN DECIMAL IS 1021


DO ANOTHER? (Y/N)
```